ENIGMAS

Arising from the 2020 Darwin College Lectures, this book presents eight essays from prominent public intellectuals on the theme of Enigmas. Each author examines this theme through the lens of their own particular area of expertise, together constituting an illuminating and diverse interdisciplinary volume.

Enigmas features contributions by professor of physics Sean M. Carroll, author Jo Marchant, writer and broadcaster Adam Rutherford, professor of earth sciences Tamsin A. Mather, professor of the history of the book Erik Kwakkel, reader in cultural history Tiffany Watt Smith, mathematician and public speaker James Grime, assistant professor of positive AI J. Derek Lomas, and explorer Albert Y.-M. Lin. This volume will appeal to anyone fascinated by puzzles and mysteries, solved and unsolved.

EMILY JOAN WARD is a medieval historian examining change and continuity between the eleventh and thirteenth centuries, and a British Academy Postdoctoral Fellow at The University of Edinburgh.

ROBIN REUVERS is a mathematical physicist interested in quantum mechanics and statistical mechanics, and a tenure-track researcher at Roma Tre University.

THE DARWIN COLLEGE LECTURES

These essays are developed from the 2020 Darwin College Lecture Series. Now in their thirty-seventh year, these popular Cambridge talks take a single theme each year. Internationally distinguished scholars, skilled as popularisers, address the theme from the point of view of eight different arts and sciences disciplines.

Subjects covered in the series include

Enigmas

Edited by *Emily Joan Ward*

The University of Edinburgh

Robin Reuvers

Roma Tre University

CAMBRIDGE
UNIVERSITY PRESS

CAMBRIDGE
UNIVERSITY PRESS

University Printing House, Cambridge CB2 8BS, United Kingdom

One Liberty Plaza, 20th Floor, New York, NY 10006, USA

477 Williamstown Road, Port Melbourne, VIC 3207, Australia

314–321, 3rd Floor, Plot 3, Splendor Forum, Jasola District Centre, New Delhi – 110025, India

103 Penang Road, #05-06/07, Visioncrest Commercial, Singapore 238467

Cambridge University Press is part of the University of Cambridge.

It furthers the University's mission by disseminating knowledge in the pursuit of education, learning, and research at the highest international levels of excellence.

www.cambridge.org
Information on this title: www.cambridge.org/9781009232548
DOI: 10.1017/9781009232517

First published 2022

Printed in the United Kingdom by TJ Books Limited, Padstow Cornwall

A catalogue record for this publication is available from the British Library.

ISBN 978-1-009-23254-8 Paperback

Contents

Figures

Notes on Contributors

Sean M. Carroll is a theoretical physicist interested in quantum mechanics, gravitation, cosmology, statistical mechanics, and foundations of physics. He is Research Professor at Caltech (California Institute of Technology), and an External Professor at the Santa Fe Institute. He is the author of several books, most recently *Something Deeply Hidden: Quantum Worlds and the Emergence of Spacetime*, and he hosts the weekly *Mindscape* podcast. He has been awarded prizes and fellowships by the National Science Foundation, NASA, the Sloan Foundation, the Packard Foundation, the American Physical Society, the American Institute of Physics, the American Association for the Advancement of Science, the Freedom From Religion Foundation, the Royal Society of London, and the Guggenheim Foundation. His TV appearances include *The Colbert Report*, PBS's *NOVA*, and *Through the Wormhole with Morgan Freeman*, and he frequently serves as a science consultant for film and television.

James Grime is a mathematician, lecturer, and public speaker. He has a PhD in mathematics and a passion for maths communication. He travels the world promoting mathematics in schools and giving public talks on the history and mathematics of codes and code breaking. He is one of the presenters on the YouTube channel Numberphile, which he helped create and which now has over three and a half million subscribers. He also has his own YouTube channel, singingbanana. He ran The Enigma Project on behalf of the Millennium Mathematics Project at the University of Cambridge.

Erik Kwakkel is Professor in the History of the Book at The University of British Columbia's School of Information. His primary interests are book design and communication in the premodern world, in particular how information was disseminated and consumed in the age before the invention of the printing press. He is widely recognised as an international expert in medieval manuscripts and is a member of the *Comité international de paléographie latine*.

He has published 10 monographs and edited volumes and runs the blog *Medievalbooks.nl*. Among his recent publications is *Books Before Print* (Arc Humanities Press, 2018), a textbook aimed at undergraduate and graduate teaching. His work is featured in a variety of public news outlets such as BBC World Service, CBC radio, CNN, and the *Smithsonian Journal*.

Albert Yu-Min Lin is an Associate Research Scientist at UC San Diego and an award-winning Explorer of the National Geographic Society. An engineer by training, he has spent the last decade developing and applying technologies towards the exploration of our shared humanity. This journey has taken him from the Arctic Circle to the middle of the Pacific Ocean, and from the deserts of the Middle East to the jungles of Central America. For this work, he has received numerous recognitions, including National Geographic's Adventurer of the Year, the United States Geospatial Intelligence Academic Achievement Award, and the Explorer's Club's Lowell Thomas Medal. He was the youngest ever recipient of the Nevada Medal. An avid science communicator, he has created nearly two dozen National Geographic and BBC documentary films, and currently hosts a National Geographic Channel series titled *Lost Cities with Albert Lin*.

James Derek Lomas is an assistant professor of Positive AI in the Faculty of Industrial Design Engineering at Delft University of Technology. His work deals closely with human experience as a central phenomenon in the field of design. He designs data-informed smart systems for human well-being, bringing humanist values into AI systems. As well as running a design consultancy, Playpowerlabs.com, he has produced over 40 learning and assessment games that have been played by millions of children. His work has received international recognition in the form of a Poptech Social Innovation fellowship, a MacArthur Foundation 'Digital Media and Learning' award, and a Sesame Workshop and White House 'Impact' Award.

Jo Marchant is an award-winning author, science journalist, and speaker. She has written several popular science books, including *Decoding the Heavens: Solving the Mystery of the World's First Computer* and the *New York Times* bestseller *Cure: A Journey into the Science of Mind over Body* (both of which were shortlisted for the Royal Society science books prize), and most recently *The Human Cosmos: A Secret History of the Stars*. Her writing explores the nature of humanity and our universe, from the mind–body connection to the mysteries of past civilisations and the awesome power of the night sky. She has a PhD in genetics and medical microbiology from St

Bartholomew's Hospital Medical College in London. She has worked as a senior editor at *New Scientist* and at *Nature*, and her articles have appeared in publications including *The New York Times*, *The Guardian*, and *Smithsonian* magazine.

Tamsin A. Mather is a volcanologist and Professor of Earth Sciences at the University of Oxford, where she has been on the faculty since 2006. She has Master's degrees in Chemistry and History and Philosophy of Science from the University of Cambridge. After a year working in Germany and Brussels doing a placement for the European Commission, she returned to Cambridge to complete a PhD on the atmospheric chemistry of volcanic plumes and their environmental effects. Since then, her research has broadened to explore the diverse ways in which volcanoes interact with Earth's environment, the processes driving volcanic unrest and eruptions, the hazards they pose, and their resource potential. Before joining Oxford, she was a Research Council Fellow at the UK Parliamentary Office of Science and Technology and a Royal Society Dorothy Hodgkin Research Fellow. She won a UNESCO/L'Oréal UK & Ireland Women in Science award in 2008, won the Philip Leverhulme prize in 2010, was UK Mineralogical Society Distinguished Lecturer in 2015/16, was the winner of the 2018 Rosalind Franklin Award and Lecture from the Royal Society, and was elected to the Academia Europaea in 2021. She has spoken at numerous science festivals, including New Scientist Live and the Cheltenham Science Festival, and has participated in several TV and radio programmes and documentaries, including Radio 4's *Life Scientific* with Jim Al-Khalili and *The Infinite Monkey Cage* with Brian Cox.

Robin Reuvers is a mathematical physicist interested in quantum mechanics and statistical mechanics. He completed his PhD at the University of Copenhagen and is currently a tenure-track researcher at Roma Tre University. While he was a Research Fellow at Darwin College, he held a Royal Society Newton International Fellowship.

Adam Rutherford is a scientist, writer, and broadcaster. He studied genetics at University College London (UCL), where he received his PhD on the genetics of the developing eye. He has written and presented many award-winning series and programmes for the BBC, including the flagship weekly BBC Radio 4 programme *Inside Science* and the long-running *The Curious Cases of Rutherford & Fry* with Dr Hannah Fry. He is the author of several books about evolution and genetics, including *A Brief History of Everyone Who*

Ever Lived and *The Book of Humans*. His most recent book, *How to Argue with a Racist* (which was a *Sunday Times* bestseller and BBC Radio 4's Book of the Week), shows how science and history can be powerful allies against bigotry. He is now an Honorary Fellow at UCL and teaches on the history of eugenics, race science, genetics, and science communication.

Emily Joan Ward is a medieval historian and British Academy Postdoctoral Fellow at The University of Edinburgh. She completed her PhD at the University of Cambridge and previously held a Moses and Mary Finley Research Fellowship at Darwin College before beginning her British Academy fellowship at University College London. Her comparative historical work examines change and continuity between the eleventh and thirteenth centuries, focusing especially on childhood and adolescence, rulership and authority, and documentary culture and historical writing. Her book *Royal Childhood and Child Kingship: Boy Kings in England, Scotland, France and Germany, c. 1050–1262* is forthcoming with Cambridge University Press. She has published several articles and co-edited *Conquests in Eleventh-Century England: 1016, 1066* (Boydell & Brewer, 2020) with Dr Laura Ashe (Worcester College, Oxford). Her passion for history communication can be heard on podcasts for *BBC History Extra Magazine* and Dan Snow's History Hit, and she has appeared in the History Hit documentary *1066: The Year of Conquest.*

Tiffany Watt Smith is the author of three books about the history of emotions: *Schadenfreude* (2018), *The Book of Human Emotions* (2015), which tells the stories of 154 emotions from around the world, and *On Flinching* (2013). Her books have so far been translated into 10 languages. Educated at the universities of Cambridge and London, she is now based at Queen Mary University of London, where she is a Reader in Cultural History and Director of the Centre for the History of Emotions. In 2019 she was awarded a Philip Leverhulme Prize for her research. Her writing has appeared in *The Guardian*, *BBC News Magazine*, and *The New Scientist*, among others, and her TED talk 'The History of Human Emotions' has been viewed more than four million times.

Acknowledgements

Both the publication of this volume and the organisation of the lecture series upon which it is based would have been impossible without the help of a wide range of people. We owe our heartfelt thanks to all those working behind the scenes to ensure the events were a success: the technicians and caretakers at Lady Margaret Hall; the fellows and students who volunteered their time as ushers and helped entertain speakers both before and after their talks; members of the Education and Research Committee; and Darwin College's incredible catering team, porters, and housekeeping and administrative staff.

There are several individuals at Darwin College whose assistance has been indispensable in guiding us through the process from the earliest stages of choosing a topic for the lecture series through to the collation and editing of this book. We are immensely grateful both to the former Master, Professor Mary Fowler, for her enthusiastic engagement with the lectures and speakers, and to the current Master, Dr Michael Rands, for his continued support for the volume since his installation in October 2020. None of this would have been possible without Janet Gibson, the College's Registrar, who tirelessly answered our emails and expertly chaperoned us through the various stages of inviting speakers, organising lectures, collating chapters, and gathering image permissions. We are indebted to Janet for all her patience, sage advice, careful management, and sense of humour while working together. Especial thanks must go to Torsten Krude and Andy Fabian for initially suggesting this volume's theme, as well as for their constant encouragement for the series. Thanks also to Espen Koht for all his IT and audio support, and to Tony Cox, Sonia Pascoal, Iosifina Foskolou, and Roger Whitehead for their help with the lectures.

It has been a pleasure to work with our eight contributors, and we would like to thank them for their involvement in this project, as well as for their patience in seeing their hard work come to fruition. We are also grateful to everyone at Cambridge University Press who enabled and supported the production and publication of this volume.

A final thanks must go to the hundreds of people who attended the lectures in person and the many thousands who watched them online afterwards. It was a privilege to share in both your excitement for our theme and your enjoyment of the lectures, and we can only hope this volume provides something of the same pleasure.

Introduction

EMILY JOAN WARD AND ROBIN REUVERS

Whether in the form of hidden codes, ancient mysteries, or perplexing conundrums, enigmas astonish and confound, but they also entice. Enigmas challenge us to ask new questions and explore novel angles of analysis. They are at the core of processes of enquiry and vital to humanity's attempts to understand the world and our place in it.

Accepting that there is no definitive answer is a humbling but often essential component of research and, indeed, of human life. The writer Douglas Adams famously satirised humanity's incessant search for conclusive answers in *The Hitchhiker's Guide to the Galaxy*. After having spent millions of years contemplating The Ultimate Question of Life, the Universe, and Everything, the supernatural computer Deep Thought eventually concludes the Ultimate Answer to be '42'. Pushed further, Deep Thought defends this answer, stating 'I think the problem, to be quite honest with you, is that you've never actually known what the question was.' As Adams's parody implies, asking the right questions can be crucial when dealing with enigmas. This is part of what makes these puzzles so compelling, the hope that revisiting an enduring problem from a unique angle or engaging with diverse perspectives may lead to fresh insights.

That hope may well be what motivated physicist Richard Feynman to suggest 'all good theoretical physicists put [137.0359...] up on their wall and worry about it'.[1] The number mentioned is the inverse of the fine-structure constant, which governs the strength of the interactions between, for example, light and charged particles such as electrons. As Feynman put it, 'Immediately you would like to know where this

---

[1] R. P. Feynman, *QED: The Strange Theory of Light and Matter* (Princeton, NJ: Princeton University Press, 1985).

number for a coupling comes from … Nobody knows. It's one of the greatest damn mysteries of physics: a magic number that comes to us with no understanding by man.' A real-life rival of 42?

Since antiquity, wisdom has been handed down in the form of enigmas, as riddles and puzzles which defied easy interpretation. The word itself derives from the Latin *ænigma* and Greek *aínigma*, both meaning 'riddle'. Its etymological roots can be traced to the idea of speaking allusively or obscurely (*ainíssesthai*) and the concept of a fable or proverb (*aînos*).[2] In the late seventh century, Aldhelm (d. 709/710), the abbot of Malmesbury in Wiltshire and then bishop at Sherborne in Dorset, amassed 100 riddles in a collection called *Enigmata*.[3] Many of his riddles are still enjoyable brainteasers today, as in the following example (the answer to which can be found in the footnote):

> From two materials, palms moulded me.
> My insides glow; these guts – for sure a looting
> Of flax or some thin reed – shine brilliantly,
> Though flesh produced from flowers yellows now.
> They're belching fire as flames and sparks are shooting,
> And maudlin tears keep dripping down my brow,
> So I still clear night's shadows that I feared;
> They leave ash smudges where my guts were seared.[4]

Aldhelm's collection, which circulated widely throughout the Middle Ages, demonstrates the persisting intrigue of enigmas as well as their power to delight. Five centuries after Aldhelm's death, another monk writing from the same abbey, William of Malmesbury, praised his predecessor's work as 'a display of sport combined with artfulness, and eloquent and vigorous expression accompanied matter of little import'.[5] Some of the enigmas discussed in this volume are, like Aldhelm's riddles,

---

[2] T. F. Hoad (ed.), *The Concise Oxford Dictionary of English Etymology* (Oxford: Oxford University Press, 1996).

[3] See, for example, London, British Library, Royal MS 12 C xxiii, www.bl.uk/collection-items/aldhelms-riddles.

[4] *Saint Aldhelm's Riddles*, trans. A. M. Juster (Toronto: Toronto University Press, 2015), no. 52, p. 31. Answer: candle.

[5] William of Malmesbury, *Gesta Pontificum Anglorum: The History of the English Bishops*, 2 vols., ed. and trans. Michael Winterbottom and Rodney Thomson (Oxford: Oxford University Press, 2007), vol. I, p. 519.

deliberate creations of human minds. Others, like the volcanoes which form the subject of Tamsin A. Mather's chapter, exist on a timescale which surpasses humanity's history. What they all share is the same propensity to fascinate, encouraging us to investigate further, hopeful of shedding further light on obscure problems or of solving difficult puzzles.

Historians often use the adjective 'enigmatic' to describe sources of uncertain provenance, perplexing content, or ambiguous meaning and purpose. One of the most well-known medieval enigmas is the fifteenth-century codex known as the 'Voynich Manuscript'.[6] Written in an unknown script accompanied by striking images, the manuscript has yet to be deciphered, despite the best efforts of experts in the century since Wilfrid Voynich discovered the text in 1912. The Holy Grail is another legendary enigma from the Middle Ages, first appearing in an Old French verse romance written by Chrétien de Troyes in the late twelfth century and infamous for its association with the mythical King Arthur and his court.[7] The Grail provides an apt metaphor for the methodology of approaching enigmatic problems. A mysterious object attracts attention, inspiring quests to discover it and often leading those seeking it down new paths on unexpected adventures. The eight chapters which follow reveal something of the innovative methodologies which are often required to uncover more about mysterious items and subjects. Some of the authors even share a little of the personal journeys (or quests!) which were a crucial part of the process of research, experimentation, and discovery.

Enigmas resonate with the processes and methodologies of research practices across the arts, sciences, and humanities. The topic's applicability across different disciplines made it an ideal theme for the Darwin College Lecture Series convened at the University of Cambridge between January and March 2020. The chapters in this volume derive from that lecture series. Despite the role of 'the enigmatic' as our contributors' guiding light, we were still surprised by the way in which the eight essays have naturally ended up speaking to each other. The authors cover

---

[6] Raymond Clemens (ed.), *The Voynich Manuscript*, with an introduction by Deborah Harkness (New Haven, CT: Yale University Press, 2016).

[7] Chrétien de Troyes, *Arthurian Romances*, trans. W. W. Kibler and Carleton W. Carroll (Harmondsworth: Penguin, 1991).

a range of concepts, problems, and phenomena to explore the idea of enigmas from multi-disciplinary perspectives. They draw out several key themes in how researchers have approached puzzles and mysteries and what these can tell us about humanity and the world around us. Together, these chapters show that investigating and resolving enigmas opens new possibilities for interpreting the physical world, the material records of the past, and even our own emotions.

A fundamental paradox lies at the heart of human life: humanity considers itself unique despite biological, genetic, and evolutionary evidence to the contrary. In Chapter 1, Adam Rutherford explains how many of the traits and behaviours once considered exclusively human are, in fact, common beyond our species. We are not the only genus that communicates. Animals make use of a range of tools, even fire. Several other species have sex for reasons other than procreation. Is humanity really that enigmatic in comparison with other organisms, especially when our existence relies on a shared code – DNA? Rutherford argues that cultural accumulators, and especially the sharing of ideas, are crucial pieces in the puzzle of humanity's evolutionary distinctiveness.

Enigmas of humanity provide a linking theme across several of the essays. Albert Yu-Min Lin, in collaboration with Derek Lomas, returns to the paradox of humanity's uniqueness as a species in Chapter 8, which concludes the volume. Focusing on human capacity to design, Lin and Lomas draw attention to the fascinating role imagination can play in human life. The relationship between human consciousness and the evolution of the species continues to captivate and puzzle scholars. In the initial lecture of the very first Darwin College Lecture Series in 1977, the philosopher Karl Popper sought to unravel some of the mysteries of this subject by discussing 'Natural Selection and the Emergence of Mind'. By revisiting the dialogue between consciousness and evolution, Lin and Lomas demonstrate how enigmas often necessitate dynamic collaboration between sciences, arts, and humanities. Archaeology provides evidence that the drive for diverse conscious experiences is no new phenomenon, while neuroscience illuminates the ways in which altered states of consciousness can enhance the variety of mental experience. Possible regional diversities in human imagination indicated by ethnobotanical and anthropological observation are placed alongside neuroimaging techniques, such as hyperscanning and biosensing, which reveal the brain's rhythmic nature. Art, design, and cognitive technologies can build on this picture by providing innovative ways of exploring conscious experience. Inspired by insights from a range of academic disciplines and reflecting on personal experience, this chapter proposes the role of 'harmony' as another enigmatic angle of research with potential to shed further light on the functioning of both human society and the human mind.

An important distinction can be made between puzzles which require enigmatic solutions and 'true mysteries', as Sean M. Carroll points out in Chapter 2. His chapter guides the reader through three related enigmas of modern physics. The first is a mystery of quantum mechanics. Despite the paradigm of quantum mechanics being pivotal to modern physics since it was put into place around 1927, important aspects of it are still not truly understood. Competing theories have been proposed, including the Many-Worlds approach, which Carroll argues fits well with other ideas at the forefront of modern physics. The second enigma is the emergence of spacetime, especially the way it interacts with gravity. Rather than following the traditional methodology of 'quantising' classical theories, Carroll proposes an alternative approach and instead seeks gravity within quantum mechanics. The chapter concludes with a discussion of the mystery of the arrow of time: what distinguishes the past from the future? Together, these three mysteries of modern physics serve as an important reminder of the endurance of enigmas in the very foundations of scholarly fields. Re-examining founding principles can provide a constructive alternative means of investigating mysteries, not only in modern science but also across other disciplines.

Enigmas not only occupy scholars in universities and academia; they also entice and fascinate researchers across a wide range of other spheres such as journalism, technology, and business. In Chapter 3, Jo Marchant shows how collaboration between different fields can sometimes prove the most fruitful way of unravelling mysteries. The chapter focuses on the fragments of a unique machine surviving from around 70–60 BC, known as the Antikythera mechanism. The efforts to decode this mystery extend more than a century since its discovery among the finds excavated by sponge divers from an ancient shipwreck off the coast of the island of Antikythera. Although the first experts who looked at the device were baffled by its gear mechanisms, dating, and purpose, Marchant explains how many of these inscrutable aspects slowly came to be clarified and solved. The initial process of deciphering the device's context and functions relied on a historian of science's fascination with ancient technology, an epigrapher's command of Greek inscriptions, and a radiographer and their X-ray machine. In the 1980s, the partnership between an amateur mechanic and a historian of computing enabled more data on the mechanism to be gathered from images taken with a crude tomography machine. But it was not until the early 2000s, when a filmmaker convinced a multinational information technology company to lend its support, that the device's remaining inscriptions could finally be read more fully. As well as illustrating the immense efforts it can take to 'solve' an enigma, Marchant's chapter displays the valuable insights which can come from these endeavours. The process of decoding the Antikythera machine challenged common assumptions about technological skill and astronomical knowledge in antiquity, but it also encouraged innovations in modern technology and revealed something of humanity's search to understand the cosmos.

Humans are skilled in producing enigmas. Sometimes these are ancient mysteries unintentionally left behind in the historical record but, as Chapter 4 shows, humans also deliberately engineer enigmas to serve their own purposes. Moving to the historical context of the twentieth century, James Grime reminds us that puzzles and codes have multifaceted uses in practices of concealment, especially for militaristic purposes, corporate secrecy, or national security. The term 'Enigma' is perhaps most recognisable in modern history and contemporary culture as the name of a cipher device used by the German military to send messages during World War II. Although the Nazis believed that the code produced by the Enigma machine was unbreakable, it was eventually cracked through the efforts of Polish and British code breakers. Focusing on the mathematics underpinning the story of the Enigma machine, this chapter sets out the process both of the code's creation and of its decryption. Grime recounts how Alan Turing, a mathematician and one of the code breakers at Bletchley Park, built on Polish efforts in the 1930s to help solve the cryptic puzzle. Turing assisted in the invention and design of the 'bombe machine' which, through a process of mathematical elimination, was able to break the Enigma machine's code. In this chapter, we see, once again, the necessity of collaborative labours when attempting to make sense of the most challenging problems. Yet Turing's story also emphasises that individual research and investigation can make a significant contribution by resolving crucial pieces of a much larger puzzle.

Describing someone as an enigma is often intended as a compliment, yet a mysterious facade is not solely an enthralling social quality. In Chapter 5, Tiffany Watt Smith focuses on the enigmatic nature of emotions and considers several different scenarios in which the act of deliberately hiding one's feelings can become an intentional strategy of defiance or defence. Although twenty-first-century social and cultural norms habitually expect a level of emotional legibility, detecting and interpreting others' emotions is never a straightforward task. By challenging some prevalent assumptions about emotions and their detection – such as the reliance on facial recognition algorithms or the belief that emotions are universal – Watt Smith judiciously reminds us that simplification is rarely the solution to an enigmatic phenomenon. Emotions are far more complex and entangled than has often been appreciated. They are also shaped by pervasive and often invisible cultural and political forces, as the first half of the chapter shows with reference to anthropological and historical examples. The second half of the chapter provides detailed examples in which the art of being emotionally enigmatic is employed as a deliberate tactic to provide protection in the face of hostility.

That we can observe enigmas even in seemingly quotidian situations, whether emotional interactions or the practice of writing, is equally relevant to Erik Kwakkel's study of medieval letter forms in Chapter 6. As Kwakkel points out, the process of designing and producing medieval books preserved historical information in both apparent and enigmatic forms. The words written by medieval scribes were, primarily, a method of conveying an evident meaning, but the shapes of the letters themselves also contained far subtler, hidden information. One of the central puzzles often faced by palaeographers is how to corroborate conclusions concerning a manuscript's provenance: how old is it, and where was it produced? While quantitative codicological studies, focused on the design aspects of medieval books, have provided indispensable insights to aid scholars in answering such questions, quantitative palaeographical studies are much rarer. Yet Kwakkel shows that studying elements of the configuration of medieval letters, especially stroke variability, can provide a new way of approaching some of the conundrums at the very centre of our understanding of medieval books. This alternative, quantitative approach to medieval letters challenges time-honoured principles of mapping medieval script families, but it also has the potential to make some of the most puzzling features of the development of medieval letters a little less enigmatic.

Chapter 7 takes us from enigmas in the everyday to an extraordinary natural object: volcanoes. Tamsin A. Mather examines some of the most enigmatic features of volcanoes and the roles they have played in the evolution of life on Earth. Valuable information can be collected from volcanoes such as Masaya in Nicaragua, which secretes gas emissions from its vents. When analysed, these data provide insights into the inner workings of our planet's hidden centre. Mather's chapter first considers how volcanic events have shaped the physical environment around us, providing examples of large igneous provinces, like the Columbia River flood basalt produced by layers of lava flowing over a huge geographical area 17 million years ago. These remarkable events have also contributed to Earth's atmosphere and climate, as more recent volcanic eruptions show in miniature. Trying to scale up present-day examples to help understand the impact of events which happened millions of years ago is one of the challenges of studying volcanism. Mather explains some of the ways researchers today are tackling this puzzle. Like many of the other chapters in the volume, this is an evocative story of how an enigmatic subject – whether a natural structure such as a volcano, an intriguing item such as the Antikythera mechanism, or a theoretical question underpinning an entire discipline – can capture the mind and inspire years of research in the hope of uncovering more about the world around us.

The 2020 Darwin College Lecture Series coincided with the onset of the Covid-19 pandemic across the world and its arrival in the United Kingdom. The eighth and final lecture of the series took place on Friday 6 March, only a few weeks before the first UK-wide lockdown came into force. The pandemic demonstrates, in all too stark reality, the negative impact of an enigmatic problem: a virus about which little was known spread rapidly throughout the global population. This exceptional event quickly altered day-to-day existence for many across the world and leaves a lasting impression on communities and individuals. Responses to the crisis share many of the features the authors of this volume have considered when addressing past and present enigmas. The search for a vaccine required new questions to be formed and deliberated as well as the design and production of innovative technologies. Collaboration across disciplines and between fields has been necessary to understand the full impact of this virus, which has had a range of physical, psychological, economic, social, and cultural consequences. And the extraordinary nature of this problem has permeated even the everyday. Although some enigmas can, at first glance, seem insurmountable in the challenges they pose to humanity, the overall impression provided by this volume is one of hope. Even problems that initially appear overwhelming can be scrutinised, interpreted, and ultimately resolved.

1 Human Origins

ADAM RUTHERFORD[1]

> 'What a piece of work is a man!' marvels Hamlet, in awe at our specialness:
> 'How noble in reason! How infinite in faculty!
> In form, in moving, how express and admirable!
> In action how like an angel!
> In apprehension how like a god! The beauty of the world!
> The paragon of animals!'[2]

The paragon of animals is a prescient phrase. Two and a half centuries after Shakespeare wrote those words, Charles Darwin would irrefutably cement humankind's position as an animal – the slightest of twigs on a single, incomprehensibly baffling family tree that encompasses four billion years and 100 billion species, all of them – of *us* – rooted in a single origin, with a shared code that underwrites our existence. The molecules of life are universally shared, the mechanisms by which we got here are the same: genes, DNA, proteins, metabolism, natural selection.

Hamlet then ponders the paradox at the heart of humankind: 'What is this quintessence of dust?' We are special, but we are also merely matter. We are animals, yet we behave like angels, or even gods. Another version of this sentiment comes from (an arguably less highbrow example of) modern culture, the Pixar animated superhero film *The Incredibles*: 'Everyone is special . . . which is another way of saying that no one is.'

This paradox is at the root of who we are. The question of what makes humans special, or even whether we are exceptional, has preoccupied our species for thousands of years, especially in the major works of most

---

[1] For further discussion of the topics covered in this chapter, see Adam Rutherford, *The Book of Humans: The Story of How We Became Us* (London: Weidenfeld & Nicolson, 2018).

[2] William Shakespeare, *Hamlet*, Act II Scene II.

religions and philosophies. In the modern age, our popular discourse is crammed with what we might call uniqueness theories, proposals of specific traits that elevated humans above and away from our ape ancestors. These suggestions cover a broad range of ideas. Some are scientifically interesting, such as our incomparable yet quintessential speech and language capabilities,[3] or our command of commodities such as tools or fire. Others are something more in the realm of speculation or assertion. In the last few years both our interest in hallucinogenic drugs and our fear of snakes have been described in the academic literature as essential drivers towards the suite of characteristics we call behavioural modernity, that is, how we became who we are today.[4]

Darwin complicated matters. We have enjoyed our current physical form for more than 3,000 centuries. We've traversed oceans of time largely physically unchanged, apart from the minuscule percentage of DNA that spells out the differences between individuals. For half a million years, our ape ancestors and cousins differed physically in degree but little in kind. From the DNA extracted from the bones of people who have been dead for thousands of years, we now know that the Neanderthals – *Homo neanderthalensis* – repeatedly interbred with *Homo sapiens*, which makes them our ancestors rather than our evolutionary cousins. People of primarily European descent carry their heritage in our genes, and people with other global ancestries bear the legacy of sexual encounters with other types of humans that no longer walk the Earth, such as the Denisovans. In 2009, a distal tip of the fifth finger of a teenage girl was found in a cave in Siberia along with an unusually large tooth.[5] That is not enough physical remains to designate a new species, a process which is still largely based on comparative morphology and not much else. But these human remains were enough to extract the full genome out of the finger bone, which indicated that this individual was neither

---

[3] For a review, see S. E. Fisher, 'Evolution of language: Lessons from the genome', *Psychonomic Bulletin and Review*, 24 (2017), 34–40.

[4] For a consideration of hallucinogenic drugs and human consciousness, see Chapter 8 by Albert Y.-M. Lin and J. Derek Lomas in this volume.

[5] D. Reich, R. E. Green, M. Kircher, J. Krause, N. Patterson et al., 'Genetic history of an archaic hominin group from Denisova Cave in Siberia', *Nature*, 468 (2010), 1053–1060.

Neanderthal nor *Homo sapiens* but was something else. We refer to those people as the Denisovans. Despite only being known from a handful of fossils, their remains harboured an entire genome that we have been able to access in the twenty-first century.

The answer to the question of what makes us human is simple: it is having two human parents. A similarly bland answer is that having a human genome is definitionally human. Neither of these factually accurate answers gets to the nub of the real questions over which we obsess. What makes us special? What is it about the human condition which is unique? *Is* it unique?

Darwin's second greatest work, *The Descent of Man*[6] – which celebrated its 150th anniversary in 2021 – largely focuses on the question that he sets up in *The Origin of Species.* In that book, in my opinion the most important book thus far written, Darwin outlines his mechanism for evolution and the radiation and diversification of organisms around the world. But he studiously avoids us. Darwin's only reference to humans in that work is a brief acknowledgement that, with his big idea in place, light will be shed upon our own species at some point in the future. It is the most tantalising 'to be continued'.

The modern era of genetics has comprehensively dealt with that tease. Though there are infinitely many questions still to be answered in the domain of human evolution, we are relatively confident about the broad sweep of humankind's last million years or so. Biologists aften state that we are an African species without much qualification. What this means is that the evidence compellingly indicates that most of our evolution occurred on that continent. *Homo sapiens* emerged from earlier apes at some point over the last half a million years or so, and for many years the focus of that evolution has been in the east of Africa, in locations around the Rift Valley. The most recent analyses suggest, however, that the oldest remains of *Homo sapiens* can be found in the northwest of Africa, from people who died some 315,000 years ago. The specimens at Jebel Irhoud in Morocco were discovered in the 1960s but, when they were

---

[6] C. Darwin, *The Descent of Man, and Selection in Relation to Sex*, 2 vols. (New York, NY: Appleton, 1871).

re-dated in 2017, they were found to be more than 100,000 years older than the oldest previous known example of *Homo sapiens*, from Omo Kibish in Ethiopia.[7]

The story that is beginning to emerge is that *Homo sapiens* represents a kind of pan-African species, and that the idea of a single location for the origin of our species is antithetical to the evidence. Nevertheless, the obviously complex picture of human evolution is often conveyed in the public sphere with simple linear narratives. These misrepresent how evolution works and what we know about the story of humankind. Among the most common examples of this are versions of the iconic image known as the *March of Progress* (which was even used to advertise the very lecture from which this chapter derives!). The original image was called *The Road to Homo sapiens*. It came from the Time-Life series of books called *Life Nature Library*, and specifically from a fold-out spread in a 1965 edition called *Early Man*. In the original version, 15 hominids progress from Pliopithecus on the far left – an ancient monkey that may well have been an ancestor of present-day gibbons – all the way to *Homo sapiens*, via Rhodesian Man, Australopithecus, Neanderthals, Cro-Magnon man, and a few others. Neither the outdated terminology used in the picture, nor the scientific inaccuracies are the problem in this image. Names change as we adopt more precise classifications, and this is to be expected as science advances. But there are two aspects of how this picture is frequently presented that are antithetical to human evolutionary thought and theory. The first is that it implies progress. The lineage depicts a road from an early gibbon-like monkey, quadrupedal, tailed, and brachiate, to a bipedal, white, modern man carrying a spear. At every stage, the move is towards a destination of an ever more upright individual, with a bigger brain, hence implying a direct, linear progression to sophistication. Evolution has no such direction and no such foresight. This image reinforces the notion of human exceptionalism and even evolutionary destiny.

The second significant issue with the *March of Progress* image is that it implies that we know this road. An irony of the progress of

---

[7] J.-J. Hublin, A. Ben-Ncer, S. E. Bailey, S. E. Freidline, S. Neubauer et al., 'New fossils from Jebel Irhoud, Morocco and the pan-African origin of *Homo sapiens*', *Nature*, 546 (2017), 289–292.

palaeoanthropology in the last 30 years or so is that, with ever more robust and voluminous evidence, we are significantly less confident of the pathway from our ancient simiiformes ancestors to us today. We can be relatively confident about the positions in time and space of the thousands of *Homo* fossils that have been unearthed, but the relationships between many of the (sometimes somewhat arbitrary) delineations of different individuals are far from clear.

What is clear is that our evolutionary tree is far from tree-like. Instead, it is pollarded, looped, and tangled. The addition of genetics to the field of palaeoanthropology has helped clarify some specific matters, and, in doing so, reveals a more – not less – convoluted history of our species. Branches of humans that evolved in Africa and bifurcated to become the separate species of *Homo sapiens* and *Homo neanderthalensis* subsequently fused in many places in Europe. Many Europeans carry Neanderthal DNA, which means that the Neanderthals were their ancestors via what we euphemistically refer to as 'gene flow events'. Similarly, the presence of Denisovan DNA in modern populations in East Asia suggests gene flow events between their ancestors and this mysterious type of human. We also know that the eastern Neanderthals had sex with an archaic form of *Homo sapiens*, both of those branches now being extinct.

There is a strong case to be made that the longstanding use of trees to describe evolution is not particularly useful when our current under-standing of our own evolutionary trajectory is packed with loops. Furthermore, our adherence to a taxonomic classification system that attempts to describe what things are, rather than what they do, is also antithetical to scientific understanding. The Neanderthals remain classi-fied as a distinct species from *Homo sapiens*, yet they happily interbred and produced fertile offspring. The Denisovans are not considered a separate species because we do not yet have enough physical remains to fulfil the accepted criteria. But their genomes are quantifiably different from Neanderthals and *Homo sapiens* in a way that implies all three were different enough to qualify as separate species. As with the hybridisation of Neanderthals and *Homo sapiens*, the gene flow events between all three indicate that, under our current species definitions, they are the same. This is obviously a bit of a mess. The historical accuracy of the story of human evolution will improve with time, but our ability to communicate the beautiful complexity of that story is fundamentally hampered when we cling to semantics and rules that are inconsistent with the evidence.

If you were to meet the people who became the Jebel Irhoud fossils, they would be largely physically indistinguishable from many people alive today. If you were to shave them or give them a nice haircut and dress them in modern clothes, you would not be able to pick them out of a picnic on Parker's Piece on a summer's day. The variations in physical appearance that we see in the people of the world today are relatively superficial. Minor adjustments in gene frequencies account for a handful of morphological traits that are the result of a little bit of geographical adaptation, a lot of genetic drift, and a history of migration and constant mixing in the last few thousands of years.

The genetic diversity of humankind is not particularly great. Our physical appearance has remained largely static over the last quarter of a million years. But what we see from the archaeological evidence when it comes to our minds is something fundamentally different. Evidence is slight, admittedly, but, on the basis of the artefacts found so far, the appearance of objects that are indicative of minds not dissimilar to our own arrived like a bolt from the blue (at least in terms of geological or evolutionary time-scales) around 40,000 years ago. Perhaps the most significant of these is the Löwenmensch – the Lion Man – of Hohlenstein-Stadel (see Figure 1.1).

Around 40,000 years ago in the hills between Nuremberg and Munich in Swabian Germany, a woman or man sat and took a piece of ivory, a tusk from a woolly mammoth, and carefully considered that it might be the right material, shape, and size for something that they had been pondering. Though now extinct, cave lions were fierce predators at that time, posing a threat to people and to the animals that people hunted and ate. That person thought about the lions, and how formidable they are, and maybe wondered what it would be like to have the power of a lion in the body of a human. Perhaps this tribe revered the cave lions out of fear and awe. Whatever the reason, this artist took that mammoth ivory and a flint knife, and patiently carved the tusk into a mythical figure.

It is a chimaera, a fantastic beast that is made up of the parts of multiple animals. Chimaeras exist throughout all human cultures for most of history: from mermaids, fawns, or centaurs to the glorious monkey-man god Hanuman, to the Japanese snake-woman nure-onna, to the Wolpertinger,

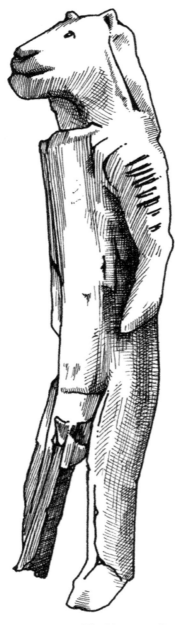

FIGURE 1.1 The Lionman of Hohlenstein-Stadel. Illustration and copyright: Alice Roberts.

an absurd and mischievous Bavarian part-duck, part-squirrel, part-rabbit with antlers and vampire teeth. The first chimaera that we know of is the Löwenmensch. It is an extraordinary piece of work, around 12 inches tall, the figure of a man with a lion's head, and an important piece in understanding our evolution. It shows profound skill, fine-motor control, and foresight in selecting the right bone and having a plan for the carving. The figure reveals an understanding of nature, and reverence towards animals in the ecosystem that affect the lives of those people. Crucially, the carving indicates a willingness to imagine a thing that does not exist.

The Löwenmensch was found deep inside the cave at Hohlenstein-Stadel in 1939, in an almost-secret vault, a kind of cubby hole which also contained other objects such as carved antlers, pendants, and beads. A few miles to the east, we find the earliest example of another charm, the Venus of Hohle Fels (see Figure 1.2). There are many examples of prehistoric sculpted female bodies. They are generically called Venus figurines, after

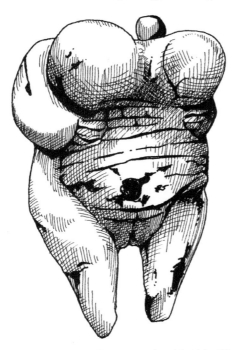

FIGURE 1.2 The Venus of Hohle Fels. Illustration and copyright: Alice Roberts. See also Figure 8.1, later in this volume, for a photograph of the Venus.

the first one discovered in the Dordogne in the 1860s by Paul Hurault, the eighth Marquis de Vibraye, who, noting the pronounced incision representing the vulva, called it *Vénus Impudique*, the 'immodest Venus'. The Venus of Hohle Fels is the most ancient of these figures, probably again around 40,000 years old. Until recently, it was the oldest known depiction of a human body.

The Hohle Fels Venus is also an abstraction. It is clearly a human, but a heavily distorted one, with features that are well beyond realism. The breasts are large, and the head is tiny. She has a huge waist and engorged labia. These enhanced sexual characteristics are also seen in some of the other Palaeolithic Venus figurines, which has led to speculation that these were fertility charms, or goddesses of reproduction. Some people have even suggested that they might be pornographic, though I think this is a bit of a stretch. We're not sure why the heads are often small, but it might be to do with perspective. Since you can't see your own head, relatively from one's own vision it is small, and, looking down, breasts may look disproportionately larger. This doesn't account for the fact that the artist could have seen the heads and bodies of other people, however. Maybe it was an artistic choice instead. If, in one million years' time, you discovered a Picasso portrait or the Bayeux tapestry isolated out of any context, you might have questions about what was on the minds of those artists. The bald truth is that we don't know the intended purpose of these beautiful artefacts, and we will never know exactly what the Palaeolithic sculptors were thinking when creating them.

What we do know is that the minds of these people were not different from our own. Though the most studied areas of the Neolithic world tend to be in Europe, this 'full package' of behavioural modernity is emerging all around the world at the same time. On a giant southern isthmus on the island of Sulawesi, part of Indonesia, there are caverns which served as the homes of people over thousands of years. About eight paces from the entrance of one particular cave is a mural measuring 1.5 metres in length. There are 12 hands – in fact, the shadows of hands, because they have been stencilled.[8] Red ochre was blown through a thin tube to outline the hands

[8] M. Aubert, A. Brumm, M. Ramli, T. Sutikna, E. W. Saptomo et al., 'Pleistocene cave art from Sulawesi, Indonesia', *Nature*, 514 (2014), 223–227.

of a long-gone person. Nearby, there is a drawing of a fat pig, and a 'pig-deer' called a babirusa. These were drawn something like 35,000 years ago, and the oldest of the handprints is 39,000 years old. In 2019, the same team who discovered these handprints also published some cave art from Sulawesi that was 44,000 years old.[9] In this image, there are multiple animals depicted in a hunting scene. This pushes the earliest figurative art back in time and around the world, away from western Europe.

Maybe this date will continue to be pushed further back in time the more we look, but, for now, behavioural modernity emerges around 40,000 years ago, in multiple locations, with little indication that it was coming. Those questions about what makes us human, or what changed to make us who we are today, are centred in this transition. The uniqueness theories that dog our discourse on this matter attempt to give singular reasons for this shift from them to us, but I find them mostly unsatisfactory. Often, behaviours that have been thought to be definitionally or uniquely human turn out to be not quite as unique as we might have thought, or not quite as informative as a pivotal human behaviour as some might hope. We do many things that today seem unconnected with our survival in a traditional evolutionary sense. We drink and play sports and go to the theatre and watch television. We devote enormous resources to enjoying sex, and yet the proportion of those encounters that result in new humans is statistically insignificant. We have decoupled sex from its primary function of reproduction. Are these behaviours unique to us? Are these the things that make us human? In what follows, I argue that many of the things we do are not nearly as unique as we might think, nor are they definitionally human. However, the way we interact and share knowledge appears to be so fundamental to humankind, and almost exclusively so, that the development of our ability to acquire and transmit culture and information is the foundation of human exceptionalism. We are, I contend, a species of teachers.

Below I will focus on two aspects of human uniqueness – or otherwise – that are of great significance to the evolution of behavioural modernity: firstly, tool use, and then, secondly, a technological subset, namely fire.

---

[9] M. Aubert, R. Lebe, A. A. Oktaviana, M. Tang, B. Burhan et al., 'Earliest hunting scene in prehistoric art', *Nature*, 576 (2019), 442–445.

Tool Use

Tools are an inherent part of our culture. Millions of years before the invention of the digital watch, we had an obligate technological culture. Scientific nomenclature even specifically acknowledges our technological commitment. One of our earliest genus cousins – probable ancestors – is named *Homo habilis*, literally 'handy man'. They were a people that lived between 2.1 and 1.5 million years ago in East Africa. The few specimens which have been classified as *habilis* generally have flatter faces than the earlier Australopithecines from around three million years ago, but still retain long arms and small heads – their brains were typically half the size of yours. To look at, *Homo habilis* would have been more ape-man than man-ape. They were probably the ancestors of the more gracile *Homo erectus*, although the two groups co-existed, perhaps indicating that *Homo habilis* diverged within its own species group.

The handy man status of *Homo habilis* is largely to do with the discovery of specimens surrounded by evidence of lithic technology. Some researchers suppose that the presence of tools represents the boundary between the genus *Homo* and what came before, meaning that humans are actually defined by tool use. The densest collections associated with *Homo habilis* come from the Olduvai Gorge in Tanzania, and this type of tech is referred to as the Oldowan tool set. It consists of chipped stone, often quartz, basalt, or obsidian, used to shape and sharpen. Obsidian is an igneous rock, a type of volcanic glass. It is a good choice for a cutting tool because it forms edges so sharp that some surgeons use it today in preference to steel scalpels.

These actions all imply a cognitive ability that enables selection of suitable stones as well as a plan. You need a hammerstone and a platform, an anvil, on which to chip away at the raw material. Knapping is a deliberate and skilful activity, and the set contains different tools. Some are heavy duty, such as the Oldowan chopper (see Figure 1.3), which we think was used as an axe head. Others are lighter-duty tools such as scrapers for removing meat from skins, or burins and tools for engraving wood. Again, this variation in the overall set of implements presupposes a cognitive ability to distinguish appropriate tools for different actions.

Homo habilis is among the earliest members of the lineage that we have decided is human, and tool use is part of that definition. This artificial

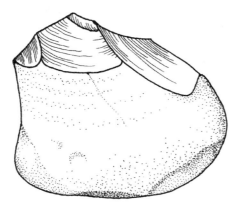

FIGURE 1.3 An Oldowan chopper. Illustration and copyright:
Alice Roberts.

boundary has not been borne out in scientific history. A thousand kilometres to the north of Olduvai is Lomekwi, on the western shore of Lake
Turkana, another one of the key areas in the nursery of early humans.
This is the site of the discovery in 1998 of a specimen which has been
designated *Kenyanthropus platyops*,[10] roughly meaning flat-faced Kenyan
man.[11] It's a not-uncontroversial earlier great ape that some have argued
is morphologically similar enough to Australopithecus that it is one, and
not a separate species. I'm not sure it matters that much since our
taxonomic definitions are very blurred at these arbitrary boundaries.
Many assumptions must be made because the specimens are few and far
between – fragments from more than 300 Australopithecine individuals
have been found, but only one *Kenyanthropus*. In 2015, a wandering team
of researchers from Stony Brook University in New York took a wrong

[10] M. G. Leakey, F. Spoor, F. H. Brown, P. N. Gathogo, C. Kiarie et al., 'New
hominin genus from eastern Africa shows diverse middle Pliocene lineages',
Nature, 410 (2001), 433–440.
[11] Historically, the word 'man' has been used to describe these species in common
parlance, as in Neanderthal man, Cro-Magnon man, etc. It's a casual usage that is
annoying in that it fails to recognise 50 per cent of our species but can be easily
corrected by generally using 'human', as in humankind, which is an easy and
inclusive fit. In this case though, 'human' specifically refers to the genus *Homo*,
which *Kenyanthropus platyops* is not in. However, -*anthropus* implies human-ness
even though in Greek it literally means 'man', so I'm not quite sure how to
represent it here. It is a hominin, which includes both *Homo* – the humans – and
Australopithecines, whose name roughly translates as 'southern ape-like things'.

turn in Lomekwi and stumbled into a site scattered on the surface with lithic detritus indicative of intentional tool-making.[12] After excavating further, they found many other fragments and tools themselves. Although not quite as sophisticated as the Oldowan set, these tools are much older, probably 3.3 million years old. In one case, a lithic flake could be matched to the stone from which it was chipped. This is viscerally powerful: imagine an ape-like person sat right there, intentionally chipping at a rock with purpose in mind. Maybe they weren't happy with how the rock split, discarded both halves, and moved onto something else. Or perhaps a ravenous predator chased them away. There, the pieces lay undisturbed for more than three million years.

We don't know who it was who sat and carved those tools, though we do know that it was a creature pre-dating the origin of the genus *Homo* – the humans – by maybe 700,000 years, and it may well have been the flat-faced Kenyans. The Oldowan tool set has now been found in key sites all over Africa where other significant evidence of human activity or presence is known, including Koobi Fora on the east side of Lake Turkana in Kenya, Swartkrans and Sterkfontein in South Africa, and further afield in France, Bulgaria, Russia, and Spain. The timescale over which this tech was used is huge, covering maybe more than a million years.

Oldowan tools, in time, were replaced with a new set of more complicated kit. Saint-Acheul is a suburb of the northern French city of Amiens where, in 1859, a major haul of axe heads would come to define the most common industry in the whole of human history. They weren't the first of these discovered – in the late eighteenth century, similar examples were found in a Suffolk village near to the market town of Diss – but they are specimens of what is now known as the Acheulean tool set.[13]

Acheulean hand axes have been worked more precisely than their Oldowan ancestors. Typically, they are teardrop-shaped, chiselled into sharp points, and crafted into flattened blades, often on both sides.

[12] S. Harmand, J. E. Lewis, C. S. Feibel, C. J. Lepre, S. Prat et al., '3.3-million-year-old stone tools from Lomekwi 3, West Turkana, Kenya', *Nature*, 521 (2015), 310–315.

[13] I. de la Torre, 'The origins of the Acheulean: Past and present perspectives on a major transition in human evolution', *Philosophical Transactions of the Royal Society B*, 371 (2016), article 20150245.

They're also larger, with a cutting edge of around 20 centimetres, compared with just 5 centimetres in a typical Oldowan blade. They represent the fruit of a concerted cognitive ability to truly craft a tool, or a weapon, and require skilled hand–eye coordination, and an even greater degree of foresight and planning. The flaking of a stone occurs in multiple stages. The initial shape is crafted, and then the blade is used to thin and sharpen the stone with a second round of delicate lithic reduction. Try it next time you're on a rocky beach, with a flint. It's a difficult, skilled process; an indelicate or badly placed blow will lethally crack the stone, and possibly also your fingers.

There is increasing symmetry in these blades as brains get bigger over evolutionary time. The instruments are found distributed around the world, and across species. The oldest Acheulean tools, as of 2015, have in fact been found in Olduvai Gorge,[14] home (at least in name) of the technology it replaced, but they are also found all over Europe and Asia. *Homo erectus* chiselled these blades, as did other early humans such as *Homo ergaster*, and Neanderthals and the first *Homo sapiens*. They used them to hunt, to butcher animals, to strip meat from skins and bones, and to carve those bones.

Acheulean tools are the dominant form of technology in human history. Though there are refinements over time, it is fascinating how stable these blades are. Many more people today use telephones or drive cars, have reading glasses or use cups, but in terms of longevity, Acheulean tools win hands down. We define this period by the technology. The Palaeolithic ranges from 2.6 million years ago until 10,000 years ago. Palaeolithic means 'old stone', which might be slightly ironic, because much of what was being engineered by those worked stones was probably wood, and bone.

Until a few decades ago, therefore, the genus *Homo* – the humans – was defined by its use of tools. But we now know that earlier apes, ones we don't call humans, were also using stone tools. Historically, we have to conclude that tool use has not been limited to humans.

[14] F. Diez-Martín, P. Sánchez Yustos, D. Uribelarrea, E. Baquedano, D. F. Mark et al., 'The origin of the Acheulean: The 1.7 million-year-old site of FLK West, Olduvai Gorge (Tanzania)', *Scientific Reports*, 5 (2016), article 17839.

This shift in perspective is borne out by examples of tool use in living non-human animals, as is discussed below. Such animals frequently use technological material harvested from trees, rather than stone, but there is no reason to suppose that early humans were not tooling wood as well. Of course, wood biodegrades, and we have scant physical remains of prehistoric tooled wood. There is a site in Tuscany in central Italy that has revealed some of the best examples of ancient carpentry. Boxwood fragments around 170,000 years old were scattered in the ground alongside Acheulean stones and bones from an extinct straight-tusked elephant, *Palaeoloxodon antiquus*. A couple of spears have been found in other sites, including in the seaside town of Clacton in Essex, but these Tuscan remains are probably multipurpose sticks, and show evidence of having been tooled, partially using fire.[15] Boxwood is hard and stiff, and the sticks show evidence of having had their bark worn away with a stone scraper, and possibly charred to remove extra fibres or knots. Who carved these spears and digging sticks? The time and place squarely put this carpentry in the hands of Neanderthals.

The Palaeolithic period covers both the Oldowan and the Acheulean tool sets, which, combined, represent more than 95 per cent of the history of human technology. Although there was a measurable shift between the two types, little else changes in the toolbox of humans for two periods of more than a million years each. There are no great leaps forward in development. Humans migrated around the world during this period, reaching all the way to Indonesia, and all over Europe and Asia. We see them slowly change in their anatomy, in species, and in their distribution around the world, but the technology remains recognisable.

Remember that these first technological humans were already maybe four million years distant from the separation of our evolutionary branches and that of chimps, bonobos, and other great apes. All of them also use tools today. What we are unsure of is the continuity of cultural tool use. Humans accumulate knowledge and skills and transmit them through time, mostly without losing those acquired abilities. Generally, we don't

---

[15] B. Aranguren, A. Revedin, N. Amico, F. Cavulli, G. Giachi et al., 'Wooden tools and fire technology in the early Neanderthal site of Poggetti Vecchi (Italy)', *Proceedings of the National Academy of Sciences*, 115 (2018), 2054–2059.

have to invent the same technology over and over again. Have all the great apes used tools continuously since that divergence, or has tool use been forgotten and re-invented many times? This is unclear, and possibly unknowable, as there is little evidence of other great apes crafting stones, even if they did use wooden tools, which are not preserved well in the fossil record. With the advent of the basic Oldowan tech in ancestors that pre-date humans – but after the split between those apes that would evolve to become us and those that would become gorillas, chimps, and orangutans – we are witnessing an ability to deliberately manipulate external objects for specific purpose that exceeds any other animal, including all the other great apes, by a significant margin.

Darwin asks whether we are unique as tool users. The answer is an unequivocal no: an estimated 1 per cent of species are obligate tool users as well, across nine different classes, including in the crustacea, within molluscs, and within birds. This indicates that tool use does not have a single origin in evolutionary history but has emerged many times as a useful functional thing. Darwin himself notes in *The Descent of Man* that gelada baboons sometimes rolled rocks down hills when attacking another species of baboon, *Papio hamadryas*. Elephants and gorillas throw stones as weapons, mostly at humans, it seems. *Lybia leptochelis*, boxer crabs, pick up and carry a pair of stinging anemones with their claws to ward off enemies, earning themselves the slightly less hardcore nickname of 'pom-pom crabs'. They'll fight other crabs if they're short of these gauntlets, and if they only have one, will rip it in half, and the anemone will grow into a cloned pair.[16]

Senegalese Fongoli chimpanzees hunt animals with weapons that they have crafted, which is rare even within the 1 per cent of animals that use tools in any capacity.[17] Once they've identified a nest of sleeping bushbabies, they find an appropriate stick, strip it, and sharpen it to a point with their teeth. They fashion it into a stabbing weapon – these spears are on average

[16] Y. Schnytzer, Y. Giman, I. Karplus, and Y. Achituv, 'Boxer crabs induce asexual reproduction of their associated sea anemones by splitting and intraspecific theft', *PeerJ*, 5 (2017), article e2954.

[17] J. D. Pruetz, P. Bertolani, K. Boyer Ontl, S. Lindshield, M. Shelley et al., 'New evidence on the tool-assisted hunting exhibited by chimpanzees (*Pan troglodytes verus*) in a savannah habitat at Fongoli, Sénégal', *Royal Society Open Science*, 2 (2015), article 140507.

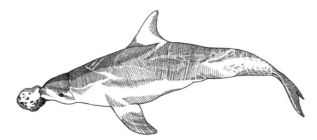

FIGURE 1.4 A sponging dolphin. Illustration and copyright: Alice Roberts.

two feet long. Bushbabies are nocturnal, and ripping open a tree cavity where they are sleeping predictably results in them scarpering. So, the specific action for the chimp is to surprise them by quickly and repeatedly thrusting the weapon into the cavity with a downward stabbing action. A rapid skewering doesn't give them that escape option. The chimps kebab the bushbabies, and eat them off the bone. This, so far, is the only example of a vertebrate other than ourselves making a tool to hunt another vertebrate.

Some bottlenose dolphins do something equally unusual as tool users: they exploit another animal as a tool. Dolphins in Shark Bay in Australia nuzzle sea sponges onto their beaks (see Figure 1.4).[18] Sponges are basal metazoans, meaning that, within the animal kingdom, they are among the least sophisticated creatures, and indeed they have no nervous system, and no brain cells at all. Around three-fifths of dolphins in Shark Bay are spongers, and researchers think the function is to protect their beaks – more technically, the rostrum – while foraging for sea urchins, crabs, and other spiky bottom dwellers that hide in the craggy seabed. The dolphins specifically select cone-shaped sponges, which presumably sit more comfortably and securely on their beaks. One animal therefore uses a second to eat a third.

Spongers therefore have a very different diet from that of non-spongers, even within the same pods. Both forage in the same areas, so we can rule out this difference being due to ecological factors – it is as if they're going to the

[18] R. A. Smolker, A. F. Richards, R. C. Connor, J. Mann, and P. Berggren, 'Sponge-carrying by Indian Ocean bottlenose dolphins: Possible tool-use by a delphinid', *Ethology*, 103 (1997), 454–465; J. Mann, M. A. Stanton, E. M. Patterson, E. J. Bienenstock, and L. O. Singh, 'Social networks reveal cultural behaviour in tool-using dolphins', *Nature Communications*, 3 (2012), article 980.

same buffet but choosing different food because one is using chopsticks. How the dolphins handle the sponge and what they eat is only a small bit of this story. The vast majority of spongers are female. They mate with males who are not spongers and have offspring, the females of which become spongers. Some animal behaviours, such as that of the boxing crabs, are presumed to be encoded in DNA, whereas other examples of tool use are acquired behaviours, built on top of a genetic and physiological frame that allows the development of that trait. The scientists who have been studying this population of dolphins since the 1980s took biopsies from spongers to establish their relatedness, which revealed something interesting. Sponging seems to stem from a single female dolphin, around 180 years ago, two or three generations back, a tool innovator now referred to as 'Sponging Eve'.[19] Because we can see the relatedness in this group, and we can see the passage of the sponging, we know that the tool use is not genetically inherited and is instead a culturally transmitted behaviour. Daughters learn sponging from their mothers, sisters, and aunts; the first known case of such cultural transmission in cetaceans.

While tool use is unequivocally not unique to humans, modern or extinct, our transmission of information about technological innovation is unusual. Though there are some examples that vaguely reflect that cultural transmission of information, such as the Shark Bay sponging dolphins, they are extremely rare. Humans, however, are cultural accumulators. We innovate and invent, and sustain that knowledge within our populations, to be passed on and built upon for generations to come. Embracing technology to extend our reach beyond our grasp has been a quintessential part of the human condition since well before the origin of our species, and the transmission of the knowledge crosses oceans of time.

Fire

There is one specific tool that is worth examining in more depth because it is paradoxically destructive: fire. The world has burnt for billions of

[19] M. Krützen, J. Mann, M. R. Heithaus, R. C. Connor, L. Bejder et al., 'Cultural transmission of tool use in bottlenose dolphins', *Proceedings of the National Academy of Sciences*, 102 (2005), 8939–8943.

years. This relentless force of nature is a chemical reaction that destroys all in its wake, from the bonds of molecules that fuel combustion, to the life that expires in the face of temperatures living cells will not tolerate. The vital molecules of biology contort and break apart, the water in our cells boils. Fire and life are incompatible.

Nevertheless, fire is part of our environment and our ecology, and the ability to adapt to, control, and use such raw power is a force that has shaped evolution. Darwin described humans' discovery of the art of making fire as 'probably the greatest, excepting language'. Maybe today we are not quite as dependent on fire as we were during the white heat of Victorian ingenuity when Darwin wrote those words, and perhaps we don't see open fires or furnaces nowadays as often as he did. But we continue to burn the energy of the sun that has been trapped within the carbon of wood, both living and so long dead it has become compressed into coal, and the energy in the carcasses of animals that perished so long ago that they have been literally pressured into becoming oil. It is in the destruction of the chemical bonds of those once vital carbon molecules that fire releases its energy. This process has shaped the modern world, and, perversely, now threatens it, as the carbon dioxide that we continue to pump out into the atmosphere itself holds more energy than other components of air and the greenhouse effect warms our world.

Fire destroys, but it also creates. Lightning strikes the ground millions of times a year, and bush fires, forest fires, and even tropical forest fires are ignited. As testament to the relentless and tenacious will to survive, plenty of plant species are pyrophytic, fire-adapted. Coniferous forests and prairies have evolved with wildfires as important drivers of ecosystem renewal. In some plants, germination can happen only after the scorching of the lands. Aleppo pines and other conifers, as well as some species of banksia in Australia, depend on the temperature of fires to melt the gum that seals their cones, and thus release their seeds, even though the heat may well kill the parent plant. Eucalyptus seeps flammable oils that encourage the spread of fire, with the effect of suppressing the creep of other plant species into their realm. In some Australian fire lilies of the genus *Cyrtanthus*, their immediate response in the aftermath of wildfires is to abruptly and prodigiously flower, advertising their desire to reproduce to eager pollinators, so they can exploit the soils newly enriched by the ashes.

Some of these plants are classified as pyrophiles – fire lovers – especially ones whose germination is dependent on the rising fire. A handful of animals are pyrophiles too. While most animals flee from the clear and present danger of combustion, the fire-chaser beetle runs towards it. It follows the infrared sensors it bears on the underside of its body; clusters of tiny liquid-filled domes expand ever so slightly in detecting radiation that is specifically given off by burning of almost any description.[20] The beetles lay their eggs in freshly burnt wood, and such is the drive to establish an evolutionary niche, and such is the sensitivity of their detectors, that they will travel 80 miles to the source of a fire to fulfil their urge to reproduce.

Humans are certainly pyrophiles too. We have had a degree of control over flames for at least 600,000 years. Over time, we moved from an opportunistic use of fire to habitual use, and eventually became obligate pyrophiles. This transition, as with all the stories in human evolution, almost certainly occurred slowly and incrementally over time. There was not one single spark, but many. Archaeologists argue over the earliest evidence of controlled use of fire but, then again, archaeologists argue about a lot of things.

We have good evidence that *Homo erectus*, that highly successful human who walked all over the Earth from 1.9 million years ago until around 140,000 years ago, was a fire user in some capacity. When this behaviour began remains disputed. Sifting through the dirt of ancient human sites is a fiddly business, and while there is molecular evidence of burnt bones and flora as long ago as 1.5 or 1.7 million years ago (depending on where you look), these are open air sites. It is not clear whether these sites are the result of wildfires or local volcanoes, rather than deliberate use by early humans. Some scholars have suggested, on the basis of shapes of teeth and other morphological dimensions, that *Homo erectus* was cooking food as long ago as 1.9 million years ago. The earliest secure date for fire in an archaeological context is probably around one million years ago in the Wonderwerk Cave in South Africa.[21]

[20] H. Schmitz and H. Bousack, 'Modelling a historic oil-tank fire allows an estimation of the sensitivity of the infrared receptors in pyrophilous *Melanophila* beetles', *PLoS ONE*, 7 (2012), article e37627.

[21] F. Berna, P. Goldberg, L. K. Horwitz, J. Brink, S. Holt et al., 'Microstratigraphic evidence of in situ fire in the Acheulean strata of Wonderwerk Cave, Northern Cape province, South Africa', *Proceedings of the National Academy of Sciences*, 109(20) (2012), E1215–E1220.

By 100,000 years ago, we largely had fire under control. As a source of heat and light, man's red fire is of obvious benefit, as is the ability not just to control it but also to use a spark to generate it. King Louie, the orangutan-in-chief from Disney's *The Jungle Book* expresses his desire to be 'just like you' by owning this uniquely human ability. He is wise to sing so. The impact of fire on the development of humankind is incomparable. We expanded north with fire as a source of heat beyond the temperate and tropical zones whence we evolved. This gave us access to a whole new range of beasts both large and small to hunt, cook, and feast upon, as well as make tools, clothes, and art from their bodies. As is the case today, the social significance of congregating around a hearth or a fire should not be underestimated. Social bonds are forged and consolidated around a fire, stories told, skills passed on, and food prepared and shared.

We are the only animal that cooks. Energy and nutrients are sometimes held deep inside the vegetation and flesh that we consume, and digestion is the process whereby they are released. This can be chemical, and mechanical. Teeth can be for grinding, tearing, and chewing, but all are for some form of maceration, the process by which food is broken down to make it more accessible to the enzymes that will chew with molecular precision. Plenty of animals use artificial mechanical means to aid digestion. Birds don't have teeth to macerate, but they do have gizzards – muscular pouches in their digestive tracts that some fill with grit that grinds up food, making it easier to chemically digest. We call these gastroliths – stomach stones – and this is an ancient practice. The fossilised remains of many dinosaurs from the Cretaceous and Jurassic periods have been found with smoothed stones inside their body cavities, where once the soft tissue of gizzards would have been.

Humans outsourced some of our digestive abilities by externalising them. By cooking foods, we break the bonds of complex molecules and make them easier to digest in our stomachs. Meat is tenderised by heating. Softer foods are quicker to eat too, in that we spend less time chewing a boiled cabbage than a raw one, which means we get access to the essential nutrients more efficiently. Dining is a period of vulnerability: when your face is occupied with ingesting a meal, it is less alert to danger from predation. Spending less time eating means less time potentially being eaten.

All these things make cooking a desirable and essential part of our evolution. Some researchers have suggested that we became a pyrophilic primate by living among iterant burning ecology and adapting to the benefits it brings. Some have suggested that the origins of cooking, or at least an understanding of how heat changes food, might have begun by apes foraging in burnt landscapes. It's difficult enough to roast a turkey to perfection in a twenty-first century oven, so it's not unreasonable to suppose that animals roasted in wildfires are most likely to be burnt or undercooked. But it may be that these first hot meals were the spark of the idea of using heat to change food for the better.

The other obvious benefit of safely standing to the side of a raging furnace is that you can be presented with an exodus of other animals fleeing from the danger. If these animals are of interest to you as food, then fire provides a free all-you-can-eat buffet. We think that South African vervet monkeys do this, enjoying unprecedented access to invertebrates scuttling out of the fire, and into their mouths.[22] We also think that the monkeys know this well and so increase their normal foraging range into wildfire regions, especially after a recent blaze. There's another set of benefits from this behaviour too. Vervet monkeys stand up on their hind legs to look out for predators, and thus can see over the grasses and plants. When they are cleared by being burnt to stubble, the monkeys can see further. Vervet monkeys in scorched plains spend more time feeding themselves and their young and less time standing erect on the look-out for something that will eat them.

Even closer relatives to us, the savannah chimpanzees in Senegal, live among fire as part of their natural ecology. It is hot in the grasslands anyway, but since 2010, the onset of the rainy season has become increasingly erratic. From October, fires start which encroach upon three-quarters of the chimps' 35-square-mile range. These most often ignite at the beginning of the rainy season, when rolling thunder and lightning meets arid bush.

---

[22] N. M. Herzog, C. Parker, E. Keefe, and K. Hawkes, 'Fire's impact on threat detection and risk perception among vervet monkeys: Implications for hominin evolution', *Journal of Human Evolution*, 145 (2020), article 102836.

Scientists have been watching these chimps for decades – the same ones that skewer the sleeping bushbabies – and in 2017 reported on their relationship with fire.[23] There are several things worth noting. The first is that they are untroubled by wildfires. Mostly they ignore the burning brush, but sometimes they wander into and explore areas that were on fire just minutes before. They appeared to navigate burnt areas frequently, which may be the same trick the vervet monkeys are using to increase their look-out range to avoid predators. In the Mara-Serengeti in Kenya, other large herbivores congregate in scorched areas at higher densities than in healthy grasslands, including zebra, warthogs, gazelles, and topi. It may also be easier and quicker for the chimps to traverse land flattened by fire.

The fact that these chimps behave in a specific predictable fashion when their world burns suggests that, while they cannot control fire, they certainly can conceptualise it, and crucially predict its behaviour. This is a cognitive benchmark, in which the animal is capable of rationalising and approaching something dangerous, rather than simply taking the safest course of action, which is to flee. The reaction is a sophisticated one: the way a fire burns is a complex and capricious process which depends on what is burning, the wind, and a host of other factors. The situation can change in a flash. Within seconds, fires reach temperatures incompatible with life, and can release smoke and noxious gases that are also threatening to apes.

The vervets and the savannah chimps are potential clues for us when we think about the genesis of our own relationship with fire. We look to nature today to draw comparisons and speculate that what we see now might be similar to what happened way back when. This may be egocentric. All data are useful in some way, but there is a whiff of presumption in the notion that behaviours in our fellow apes reflect our own journey to the present.

Is this what we did? Do the chimps today mimic our own evolution 100,000 years ago, or even a million? These are hard questions to answer. Neither bones nor the ground preserve behaviour well. But we can see

[23] J. D. Pruetz and N. M. Herzog, 'Savanna chimpanzees at Fongoli, Senegal, navigate a fire landscape', *Current Anthropology*, 58 (2017), 337–350.

how bodies change in relation to changing environments, such as the subtle shifts away from an arboreal life, and infer what behaviour was facilitated by those bodies. And we have better clues and tools to answer the question of how fire changed us, though the evidence is almost as fleeting as wisps of smoke. We look for charred remains buried in the dirt, or for evidence of hearths and kitchens. We also look at the morphology of ancient humans, to see if cooked food was a necessity for shaping their bodies, or at least the hard parts that remain for us to scrutinise today. Body mass and feeding times can help construct models of the energy required to make those bodies, allowing us to calculate the particular dietary requirements they demanded. We construct tests in our living primate cousins and see how they tally with the behaviour that we are just beginning to observe in the small pockets of monkeys and chimps which encounter fire on a regular basis.

These are data points that may build up a theory, but we should be cautious. Most great apes do not live on the savannah. The majority of chimpanzees, bonobos, gorillas, and orangutans live in dense forest environments, where blazes are only devastating and mercifully rare. There are few formal reports on the effect of forest fires on the lives of great apes, but peat burning in Indonesian national parks (which is associated with expansion of palm oil farming) has had a solely detrimental effect on orangutans. In 2006, hundreds were estimated to have died as a direct result of forest fires.

During our evolution in Africa, savannahs expanded, forests shrank, and our morphology inched away from being adapted to an arboreal life. Single causes are rarely persuasive arguments for how we evolved to be what we are today. Though our transition into the species *Homo sapiens* occurred in Africa, I think we are moving towards understanding ourselves as being a kind of hybrid derived from multiple early African humans. Certainly, although the strongest evidence comes from the east of Africa, we haven't really looked that hard over the rest of that vast landmass. The earliest known *Homo sapiens* are found in Moroccan hills to the east of Marrakesh. Fire is undoubtedly one of the great driving forces of human evolution, but it is not the only one. Our existence in the perpetual presence of savannah fires almost certainly profoundly changed us, but our ancestors didn't all live on the plains of Africa.

Darwin said that, out of all the animals, *Homo sapiens* 'alone makes use of tools or fire'. There, he is quite wrong. None bar us can ignite a fire or create a spark. However, we are not alone in using fire as a tool. Corvids are adept tool users but, until 2017, raptors – that is, birds of prey – were not known for their tool-using abilities. Raptors is an informal and broad classification, one which includes kites, eagles, osprey, buzzards, owls, and so on, and therefore doesn't necessarily relay evolutionary relatedness. Owls are closer to woodpeckers, and falcons closer to parrots than either raptor is to hawks or eagles. They are all hunters though, with curved talons and beaks, and tend to have keen eyes, some with impressive varifocal vision, honed to zoom in on a tiny mammal when soaring above.

A few of these birds are also pyrophiles. Fire-foraging raptors work on similar principles to the vervet monkeys. Tasty critters will be flushed out of a burning bush and are easy pickings. Plenty of raptors eat carrion, and there will be lots of roasted small mammals in among the ashes. This behaviour has been noted in the scientific literature as early as 1941, all over the world, including east and west Africa, Texas, Florida, Papua New Guinea, and Brazil.

Some raptors are even smarter. Black kites, whistling kites, and brown falcons all have international ranges, and are indigenous to Australia, where they hunt and scavenge carrion, particularly in the baked northern tropical savannah. These Australian lands are hot and tinder dry, and they burn regularly. Aboriginal Australians know this well and have managed the fires with great sophistication for thousands of years. They use fire to raze particular flora and encourage the growth of edible plants and grasses that attract kangaroos and emu, both of which make good meat.

The indigenous people also know the local fauna. Over several years, culminating in a study published in 2017, Aboriginal rangers and subsequently Australian scientists reported that black kites, whistling kites, and brown falcons have all been seen doing something very thoughtful.[24] These raptors pick up burning or smouldering sticks from bush fires and carry the torches away. Sometimes they drop them because they are too

---

[24] M. Bonta, R. Gosford, D. Eussen, N. Ferguson, E. Loveless et al., 'Intentional fire-spreading by "firehawk" raptors in northern Australia', *Journal of Ethnobiology*, 37 (2017), 700–718.

FIGURE 1.5 A firehawk. Illustration and copyright: Alice Roberts.

hot, but the intention is to place them in dry grassy areas and set a new blaze. Once alight, they perch on a nearby branch and await the frenzied evacuation of small animals from the inferno. Then they feast.

Aboriginal Australians have known of these fire-starters for a while.[25] They refer to the birds as firehawks (see Figure 1.5), and the raptors feature in several religious ceremonies. There is a sighting recorded in one account from *I, the Aboriginal*, the 1962 ghost-written autobiography of an indigenous man called Waipuldanya:

> I have seen a hawk pick up a smouldering stick in its claws and drop it in a fresh patch of dry grass half a mile away, then wait with its mates for the mad exodus of scorched and frightened rodents and reptiles. When that area was burnt out the process was repeated elsewhere. We call these fires Jarulan ... It is possible that our forefathers learnt this trick from the birds.[26]

---

[25] This research is led by Bob Gosford, an Australian ethno-ornithologist who lives, appropriately enough, near Darwin in the Northern Territories. They refer to indigenous ecological knowledge (IEK) and go to great and necessary lengths to acknowledge, engage with, and build on the longstanding traditions and skills of the first peoples of Australia. This is a somewhat new practice, but shows clearly how much is to be gained in the sphere of understanding our world by respecting indigenous people with humility and grace.

[26] Douglas Lockwood, *I, the Aboriginal (A Biography of Waipuldanya, Also Known as Wadjiri-Wadjiri or Philip Roberts)* (Adelaide: Rigby, 1962).

In the diaphanously thin academic literature on this incredible phenomenon, there has been some historical dispute over whether this fire-starting is deliberate or not. The most recent study in 2017 is the first formal scientific account, and it concludes from multiple eye-witness testimonies over many years that this fire-starting is fully intentional.

This, as far as I am aware, is the only documented account of deliberate fire-starting by an animal other than a human. These birds are using fire as a tool. By any of the definitions mentioned above of what constitutes a tool, this behaviour satisfies all of them. It also goes some way to explain how fire can apparently jump over human-crafted and natural fire barriers, such as barren paths or creeks. It is possible that Aboriginal Australians learnt to start *jarulan* from the birds, and later adopted it into their management of the fires that have burnt throughout Australia's history. If true, this is a beautiful example of cross-species cultural transmission. It is also possible that our ancient ancestors did the same more than a million years ago, when we began a relationship with fire that will never be extinguished. Or maybe it is just a good trick, and only we and the raptors have worked it out. Either way, the ability to start a new blaze is one of the first steps in being able to control fire.

This does not mean the next steps will follow. These hawks are not necessarily en route to forging metal or cooking food. But the birds' knowledge does move one step beyond the behaviour of the vervet monkeys and the Fongoli chimps. As well as requiring a cognitive understanding of the behaviour of fire, not least how dangerous it is, the raptors also demonstrate an ability to plan ahead and calculate a considerable risk. At what age would you let a child handle a burning stick? The falcons and kites are using a lethal force of nature to manipulate the environment for a meal that otherwise would have remained safely hidden in the bushes.

Fire is part of nature. The world has burned since before there was life. Nature, with its tenacious ability to adapt to the environment in front of it, has repeatedly embraced the inferno. We have gone a few steps further and created a total dependence on this raw power. Humans can't survive anymore without cooked food. We do have other sources of energy nowadays, but we remain utterly dependent on burning the remains of long dead animals and plants, at least for the foreseeable future. Using fire is part of our nature, and you can't start a fire without a spark. Although we are the only ones who can do that, we now know that humans are not the only species who see fire as a means of getting what we want.

Human Origins: A Shifting Puzzle

These two examples, tools and fire, indicate that some of our behaviours are not quite as unique as we once thought. Our ancestors in the Palaeolithic were biologically similar, but the question still remains: *why did it take so long to become modern when we were physically ready for thousands of years?*

There are many pieces of this puzzle that remain elusive, although blossoming areas of research, such as Theory of Mind and the nature of consciousness, are potentially promising avenues. New research resurrects questions that have languished in fascinating philosophical realms for decades and centuries but begins to examine them with more precise twenty-first-century scientific tools. As these areas become entwined with neuroscience, we inch towards a better understanding of human behaviour.

One idea that I think is crucial has been attracting increasing attention in the last few years: the relationship between changing population size and structure and the onset of modernity. How we organised our society mattered. The first clue to this theory is that populations seem to grow larger in multiple locations around the onset of modernity. We see it in Africa 40,000 years ago, and at a separate time in Australia, more like 20,000 years ago. Such expansions may be in line with the local environment: as the climate changed, life became easier. They might also be a manifestation of our huge migrations. No other creature has moved so permanently in such a short period of time. Within 20,000 years of leaving Africa, we had settled in Australia.

The opposite effect can also be observed: a loss of cultural sophistication in societies whose populations do not grow or migrate, or where they are cut off from a bigger populace. For example, Tasmania became an island around 10,000 years ago, as the last Ice Age thawed. The seas rose, and it became separated from mainland Australia by what Europeans named the Bass Straits. The indigenous people of Tasmania managed to maintain a tool kit of only 24 pieces in that isolation, losing the skills to make dozens of others over thousands of years in the Neolithic. Indigenous Australians on the mainland developed more than 120 new tools during the same period, including multi-toothed bone harpoons.

In the Tasmanian archaeological record, we see the gradual disappearance of fine bone tools, the loss of the ability to make cold weather clothing, and, perhaps most significantly, the degradation of fishing technology. Hooks and spears for catching cartilaginous fish vanish from archaeology, as does evidence of fish bones (although the population did continue to forage and eat crustaceans and sessile molluscs). When Europeans arrived in the seventeenth century, the indigenous people expressed both surprise and disgust at the colonisers' skill at catching and eating large fish, yet 5,000 years earlier it had been a key and thriving part of their diet and culture.

Scientists who are interested in a fuller understanding of humanity's development have created models to try to understand how the cultural transmission of skills is affected by a population's size and structure. In this way, they can test how and why we see the hallmarks of modern behaviour come and go, then eventually stay in the archaeological record. These are effectively equations that model how an idea or skill is passed around in a community. They plug in hypothetical numbers for the size and density of a population, and a skill level for an imagined expert task – maybe knapping an arrowhead or tooting a flute – and then they run simulations that work out how that skill level can be transferred between people. Mathematical models of this sort are pretty technical, but what they are effectively doing is saying 'Here are people with a very particular set of skills, which can be taught to others. How does the size of a population affect the efficiency of teaching?'

The answer to this question appears to be 'enormously'. Larger populations enable the transfer of complex cultural skills with far greater efficiency than smaller ones. The maintenance of skill levels is heavily dependent on population size (which is also affected by migration). According to the models, small populations, especially isolated ones, will lose skills through an inefficient transmission. When populations grow, they accumulate culture more readily. Only we do this. And we do it all the time, whereas there are only a few select examples of cultural transmission in other animals.

Humans are social, meaning that we depend upon interactions with others for our own well-being. We are cultural transmitters, passing on a wealth of knowledge that is not encoded in our DNA. This transmission is horizontal, not just vertical, meaning that we teach to people who are not our children and may not even be genetically close kin, but are our peers. Despite being highly skilled and creative, our expertise is not distributed evenly throughout our populace. Some people have skills that others don't, and when we need to find out how to do something, we ask an expert. The demographic structure of a society is essential in maximising the way information and skills are transmitted within a group. Any collection of people relies on an internal organisation to be effective. These mathematical models seem to show that our modernity – the full package of being the humans that we are today – depends on being able to accumulate culture, to pass it on, and to transmit it in a society that grew to optimise the overall success of its members.

This is territory that is actively being researched right now.[27] It is the model that I think is right, for what that is worth, though much more work is needed. A tiny proportion of the ground has been dug to uncover our pasts. A fraction of the genes of our ancestors have been sampled. As ever in science, the answers are never complete, and we mould and carve ideas, throw them away if the data don't fit, or build them up if the data do fit. The idea that demography was an essential pivot in the ascent of us is an idea that is young. The truly amazing thing is that Darwin was thinking along the very same lines, one and a half centuries ago. He writes in *The Descent of Man*:

> As man advances in civilization, and small tribes are united into larger communities, the simplest reason would tell each individual that he ought to extend his social instincts and sympathies to all members of the same nation, though personally unknown to him. This point being once reached, there is only an artificial barrier to prevent his sympathies extending to the men of all nations and races.

---

[27] I recommend the work of Mark Thomas at UCL and Joseph Heinrichs at Harvard, e.g., J. Henrich, R. Boyd, M. Derex, M. A. Kline, A. Mesoudi et al., 'Understanding cumulative cultural evolution', *Proceedings of the National Academy of Sciences*, 113 (2016), E6724–E6725.

2 Mysteries of Modern Physics

SEAN M. CARROLL

Physics has both puzzles – aspects of phenomena that are incompletely explained within the dominant paradigm – and true mysteries – background issues that aren't understood to anyone's satisfaction, but which we've largely bracketed in favour of moving on and making progress elsewhere. I will focus on the latter, picking out three of my favourite mysteries. The first is the mystery of quantum mechanics, a wildly successful empirical theory whose complete formulation is still lacking. The second is the emergence of spacetime, a conundrum that lies at the intersection of quantum mechanics and gravitation theory. And the final one is the arrow of time: what distinguishes the past from the future. We attribute the arrow to the increase of entropy over time, but the connection between entropy and all aspects of time's arrow is still mysterious, as is the reason why entropy started out so low in the first place.

Fundamental Physics, Ancient and Modern

Like any good science, physics is replete with mysteries. Scientists spend their time pushing at parts of their subject they don't understand, and once they do understand them, they move quickly on to the next issue. And like any good science, physics goes through different stages with different kinds of challenges. In Thomas Kuhn's famous formulation,[1] there are periods of 'normal' science, where researchers work at well-defined puzzles within the context of a common paradigm, and periods of 'revolutionary' science, where one paradigm is overthrown in favour of another.

---

[1] Thomas S. Kuhn, *The Structure of Scientific Revolutions* (Chicago, IL: University of Chicago Press, 2012).

Fundamental physics – the study of the very basic building blocks of nature, to the best that we currently understand what those are – is currently in a normal-science period. The various parts of fundamental physics – quantum mechanics, particle physics, gravitation, cosmology – all feature well-established 'standard models' that provide excellent fits to all the data we've currently collected. But, in each case, we have reasons to believe that these models are incomplete. In such a situation, one strategy is to try to push forward, essentially by guessing how even better theories might be constructed: that's an extremely popular approach. Another strategy is to take a step back and examine the very foundations of the field, looking for weak spots where new ideas might be useful. It's those kinds of mysteries that I want to talk about in this chapter.

To lay the groundwork, let's take a quick look at what our best ideas about physics were before the modern era. By 'modern', in this case, we mean 'before Isaac Newton in the 1600s': physics is the study of what the world is made of and how it behaves, and people certainly gave thought to that question before Newton arrived on the scene.

One important idea goes all the way back to Aristotle, two millennia before Newton. He thought about the world in terms of causes and natures. It was a teleological view of physics: different objects had different natural states, and if pushed away from those natural states, they would try to return to them. Most objects, for example, were naturally at rest. To get something to move you have to push it, and when the push disappears, the object will return to rest.

This idea is not at all crazy. If you have a coffee cup in front of you, you can get it to move by gently pushing it across the table. And when you stop pushing, the cup will stop moving. It was an enormous step forward when people first suggested that the 'natural' state of things was uniform motion, not returning to rest – that the reason why the cup stops moving is the friction caused by the table, not its natural inclination.

As far as I know, it was Ibn Sīnā (sometimes Latinised as Avicenna) who first made a concrete proposal along these lines (although to be fair, the history is murky and full of fits and starts). Ibn Sīnā was a Persian polymath in the Islamic Golden Age, around 1000 AD. He argued that, if we could do a thought experiment where objects were moving in empty space (the vacuum), they wouldn't slow down at all. These days, of course, we have direct evidence of this in the form of artificial spacecraft zooming around the Solar System.

Ibn Sīnā's insight was the beginning of a simple rule we now call 'conservation of momentum'. Every object has a certain property called its momentum (roughly its mass times its velocity), which remains constant unless we act upon that object with some force. This seems innocent enough, but it represents a huge break with the existing picture of physics as a story of causes and natures. That view gets replaced by a picture based on *patterns*, known as 'the laws of physics'.

This pattern-based view was developed into the theory now known as 'classical mechanics' by luminaries such as Galileo and Newton, but its full force wasn't appreciated until the work of Pierre-Simon Laplace in the nineteenth century. Laplace pointed out an implication of classical mechanics that today we would call 'conservation of information'. This idea was expressed in the thought experiment of Laplace's Demon: a vast intelligence with infinite calculational capacity and a perfect knowledge of the laws of physics. If you tell Laplace's Demon the positions and velocities of every particle in the universe at any one moment of time, it can predict the exact position and velocity of those particles at any other time, whether it be in the past or future. The information needed to specify the entire history of the universe is contained in every moment. This is the picture that has come to be known as the 'clockwork universe'.

Modern physics, especially quantum mechanics, has somewhat tarnished the pristine determinism of classical mechanics. But the basic idea of an underlying pattern, expressed in the laws of physics, is still with us. Our current best working model of the laws of physics was dubbed the 'Core Theory' by physicist Frank Wilczek.[2] The entirety of the theory can be summarised in a single compact equation:

[2] Frank Wilczek, *A Beautiful Question: Finding Nature's Deep Design* (London: Penguin, 2015).

$$A = \int_{k<\Lambda} [Dg][DA][D\psi][D\Phi]$$

$$\times \exp\left\{ i \int d^4x \sqrt{-g} \left[\frac{1}{16\pi G} R - \frac{1}{4} F_{\mu\nu} F^{\mu\nu} + i\bar{\psi}\gamma^\mu D_\mu \psi \right.\right.$$

$$\left.\left. + \left| D_\mu \Phi \right|^2 - V(\Phi) + \left(\bar{\psi}_L^i \Upsilon_{ij} \Phi \psi_R^j + \text{h.c.} \right) + \sum_a \mathcal{O}^{(a)}(\Lambda) \right] \right\}.$$

This looks intimidating, but, considering that it agrees with every physics experiment ever performed here on Earth, it's an impressive achievement. It would take us too far afield to explain the significance of each term in the equation, but it is the modern equivalent of what Laplace's Demon would use to evolve the state of the world from one moment to another. It includes everything we know about spacetime, quantum mechanics, particle physics, and gravity (at least in relatively mild conditions).[3]

But it doesn't include everything; physics isn't finished, or anywhere close. Once we go off the Earth and out into space, we readily encounter phenomena we cannot explain within the Core Theory. There is dark matter, a kind of invisible particle that can't be explained by any of the ingredients in the Core Theory. There is dark energy, the mysterious quantity suffusing space that drives the acceleration of the universe. There is the imbalance between matter and antimatter. There is the Big Bang itself, not to mention black holes, both of which represent extreme gravitational phenomena that lie outside the Core Theory's domain.

There are, in other words, plenty of puzzles that the Core Theory doesn't have the answer to. But as far as we know, these are normal-science puzzles, which physicists are trying to tackle within their standard paradigms. The Core Theory is essentially a placeholder, a compact statement of what we know so far, without any pretension to being complete. The fact that it agrees with experiment is therefore quite frustrating. Significant experimental anomalies provide helpful clues to how we should go beyond our current theories and construct something

---

[3] Sean M. Carroll, 'The quantum field theory on which the everyday world supervenes', *arXiv preprint* arXiv:2101.07884 (2021).

better. Lacking that, it can be useful to re-examine deep issues at the foundations of physics, which is what we turn to now.

Mystery 1: Quantum Mechanics

Remarkably successful as the clockwork universe of classical mechanics was, at the beginning of the twentieth century it crumbled under an onslaught of new data and ideas. The resulting new paradigm, put into place around 1927, is quantum mechanics. This theory is even more successful than classical mechanics. Today, quantum mechanics lies at the heart of all of modern physics, and has essentially no competitors as an underlying framework for understanding the physical world.

This makes it all the more remarkable that, in the famous words of Richard Feynman, 'I think I can safely say that nobody understands quantum mechanics.'[4] We use quantum mechanics all the time, to build incredible technological devices and to make extraordinarily accurate experimental predictions, but there is something important about it that we don't truly understand.

Quantum mechanics began when physicists were trying to understand the complementary roles of waves and particles in the fundamental building blocks of nature. In the nineteenth century, James Clerk Maxwell and others had assembled an amazingly fruitful theory of electromagnetism, which included the idea that light is an electromagnetic wave. But, at the turn of the century, Max Planck and Albert Einstein were able to explain certain puzzling experimental results by postulating that light sometimes behaves like a particle.

Meanwhile, there was a problem with atoms, or at least with the usual cartoon of electrons orbiting around a central nucleus, much like planets orbiting the Sun. The problem is that those electrons move so rapidly that they should emit radiation, lose energy, and spiral down into the nucleus, all within less than a hundred-billionth of a second. Niels Bohr patched up the problem by positing that electrons were only allowed to

---

[4] Richard P. Feynman, *The Character of Physical Law* (Cambridge, MA: MIT Press, 1967).

be in certain orbits, and would 'jump' from one to another at unpredictable intervals. But this seemed somewhat *ad hoc* and unsatisfying.

Inspired by the idea that light has particle-like properties, Louis de Broglie suggested that electrons (or matter particles in general) can also have wave-like properties. This could better explain the stability of atoms: electron waves would naturally fall into certain discrete shapes, much like the harmonics on a plucked violin string. Erwin Schrödinger promoted this idea to a full-fledged theory, together with a famous equation that bears his name.

There was only one problem with the theory that electrons are waves: they certainly *look* like particles. When you send an electron through a detector, it leaves a discrete track, like the trajectory of a particle. It doesn't look like a fuzzy, wave-like cloud. It's almost as if electrons behave differently when we are and are not observing them.

In the face of this puzzle, physicists took the bold step of simply accepting it. They proposed that quantum systems followed two different sets of rules, depending on whether they were being observed. When we're not looking at them, systems are described by 'wave functions' that obey the Schrödinger equation. It's a perfectly deterministic evolution, not that different from the classical clockwork universe. But when an observation or measurement occurs, a completely different rule kicks in. Then, the wave function 'collapses' onto some particular measurement outcome; for example, if we're observing the position of a particle, the wave function instantly changes to be localised around some particular position, even if it was diffusely spread out before. And we can't predict exactly what that position is going to be, it's fundamentally random. What we *can* calculate is the probability of different measurement outcomes: the probability is high where the original wave function was large, and low where it was small. This is known as the Copenhagen interpretation of quantum mechanics, put together by Bohr, Werner Heisenberg, Max Born, and others.

The problem, as should be obvious, is that this is completely unacceptable as a fundamental theory of nature. There are two problems in particular.

The first is the *measurement problem*. When you and I talk about measuring something, it's roughly clear what we have in mind. But when we're trying to formulate a fundamental theory of nature, a much greater level of precision is needed. What precisely do we mean by 'measure' or 'observe'? When exactly does it happen? Does it have to be a living, conscious observer? Or would a video camera do just as well? What about completely passive interactions with the environment, like the air molecules in a room?

The other issue is the *reality problem.* Quantum mechanics is all about the wave function, which we use to calculate the probability of getting particular measurement outcomes. But what *is* the wave function? Is it supposed to be a direct mathematical representation of reality, like positions and velocities are in classical mechanics? Or does it represent only a part of reality, and are there additional quantities needed to capture the whole truth? Or, most radically, is the wave function not directly related to reality, but merely a tool we use to predict experimental outcomes to the best of our ability?

We don't know. There are competing theories about what the fundamental formulation of quantum mechanics should be, and the various alternatives take all the possible positions outlined above. That's the mystery, and it's a noticeably embarrassing one for physicists: we can't agree on what our most important theory actually says.

There isn't space here to examine all the possible versions of quantum mechanics, but I want to focus on the one that is my personal favourite: the Everett, or Many-Worlds, approach to quantum theory. Hugh Everett III proposed this idea as a graduate student in the 1950s. It is extremely modest in its assumptions, but wonderfully radical in its implications.

Copenhagen quantum mechanics uses the wave function as a way of calculating probabilities, and it posits a different kind of evolution (wave function collapse) when systems are measured. Everett is more austere and straightforward. The wave function is nothing more than a direct representation of the real world; it always obeys the Schrödinger equation, without exception. This is certainly a leaner and meaner version of quantum mechanics than the conventional view, at least in terms of its basic formulation.[5]

---

[5] Sean M. Carroll, *Something Deeply Hidden: Quantum Worlds and the Emergence of Spacetime* (Boston, MA: Dutton, 2019).

The question is, how can something that simple possibly be true? We've already said that electrons don't appear to us as waves when we observe them; they look like particles, as if the wave function has collapsed. How can Everett account for those observations if there is no wave function collapse in his theory?

The answer lies in a feature of quantum mechanics called 'entanglement'. We can illustrate the idea with the infamous Schrödinger's Cat thought experiment. This was a set-up invented by Schrödinger to illustrate the apparent absurdity of the Copenhagen interpretation. (While he was one of the founders of quantum theory, Schrödinger was never completely happy with how it was ultimately formulated.) We know that microscopic quantum systems can be in superpositions of different measurement outcomes; for example, a radioactive nucleus can be in a superposition of 'decayed' and 'not decayed'. Schrödinger proposed a way to amplify such a superposition to a macroscopic system, in this case a cat that he (in his imagination) put in a superposition of 'asleep' and 'awake'. (In reality, he imagined the cat was in a superposition of 'dead' and 'alive', but there's no reason to imagine killing the cat, asleep and awake work just as well.) His question was: do you truly believe that the cat is neither awake nor asleep, but in a superposition of both, until we observe it?

Here's what Everett's interpretation would say. Imagine we put the cat into a superposition of two different alternatives. As soon as the cat interacts with something else, even just the air in the room, the state of the cat will become entangled with the state of the air. That means that the quantum state of the universe describes a superposition of two different pieces: one where the cat is awake and the air has interacted with an awake cat, and the other where the cat is asleep and the air has interacted with a sleeping cat. That's a direct and undeniable prediction of the Schrödinger equation, if you don't put in wave function collapses by hand.

Everett's major realisation was that this prediction is perfectly compatible with what we observe. The trick is to notice that the two pieces of the wave function – one with the cat awake, the other with it asleep – will *never interact with each other again.* Whatever happens in one, the other piece goes on completely unaffected. It is as if they have become separate worlds. Therefore, says Everett, when you look at the cat, you shouldn't think that you either see it awake or you see it asleep, nor should you think that you're in a superposition of both. Rather, you should think that you have branched into two distinct people, one of whom sees an awake cat while the other sees a sleeping one.

The important point here is that Everett never modified quantum mechanics to introduce multiple worlds. The worlds were always there. If you think that the wave function represents reality, and you think that a single particle (or a cat) can be in a superposition of multiple possibilities, then you should think that the universe as a whole can be in a superposition of multiple possibilities. Everett just points out that this actually happens, and that we can treat the different parts of the superposition as independently evolving copies of the universe.

So is Everett's interpretation right? We don't know. There are certainly alternatives, such as, among others, pilot-wave theories that add variables besides the wave function, objective-collapse theories that modify the Schrödinger equation with random jumps, and epistemic approaches that think of the wave function as a measure of our knowledge rather than as a representation of reality.

But there is something attractive about the Many-Worlds approach. It relies on very simple assumptions, even if the ultimate predictions seem pretty wild. And it seems to fit well with other ideas at the forefront of modern physics, including the notoriously difficult problem of quantum gravity. Let's look at that next.

Mystery 2: Spacetime and Gravity

Albert Einstein is an iconic figure, rightly chosen by *Time* magazine as its Person of the Century. In some popular discussions he is depicted as having been somewhat out of touch when it comes to quantum mechanics, too stuck in his ways to keep up with the latest developments. That impression is wildly off the mark. Einstein was one of the inventors of quantum mechanics, and he understood it as well as anyone. He was simply convinced that we didn't understand it completely yet, a conviction which our above discussion seems to support.

Einstein's biggest claim to fame is the theory of relativity, and in particular *general* relativity. 'Special' relativity came first, around 1905, when Einstein tied together a number of strands that other physicists had already proposed. His breakthrough was to understand that we didn't need to posit an invisible 'aether' through which light propagated; electromagnetic waves simply move through empty space, and they always do so at the speed of light, regardless of how the person observing them might be moving themselves. Einstein's old teacher Hermann Minkowski later realised that the best way to think about this insight was to combine space and time together in a single, four-dimensional spacetime.

But spacetime in special relativity is simply a static background, an arena in which the rest of physics plays out. And, for various technical reasons, it was difficult to fit the theory of gravity into this framework. It wasn't until a decade later that Einstein finally resolved these issues with the theory of general relativity. He suggested that spacetime is not a fixed background – it's a dynamical, changing entity in its own right, responding to the presence of matter and energy. The important feature of spacetime is its curved geometry, which can deviate from the rectilinear rules handed down by Euclid. This deviation is what you and I perceive as gravity. Objects appear to be pushed around by gravitational forces because they are responding to the curvature of the universe around them.

General relativity takes the idea that gravity is the curvature of spacetime and makes it tremendously quantitative. The equation that Einstein wrote down in 1915, relating spacetime curvature to the energy of matter and radiation and other sources, has never been contradicted by experiment. Like any really good equation, it knows a lot more than the person who first wrote it down: it implies the existence of black holes, gravitational waves, and the Big Bang, none of which Einstein himself knew at the time. All these predictions have been wonderfully confirmed by experiment, for example the recent discovery by the LIGO and VIRGO observatories of gravitational waves produced by colliding black holes.[6]

---

[6] B. P. Abbott, R. Abbott, T. D. Abbott, M. R. Abernathy, F. Acernese et al. (LIGO Scientific Collaboration and Virgo Collaboration), 'Observation of gravitational waves from a binary black hole merger', *Physical Review Letters*, 116 (2016), article 061102.

Unfortunately, general relativity doesn't seem to play nicely with the other great pillar of modern physics, quantum mechanics. Aside from the mysteries of the quantum measurement problem and the reality problem, physicists have been very successful at taking classical theories and 'quantising' them to develop functioning models of the quantum world. This works fantastically well with all the known matter particles and force fields, except for gravity. (The Core Theory, mentioned above, includes gravity, but only in the 'weak field' limit. It doesn't describe extreme phenomena like black holes or the Big Bang.)

Physicists have therefore been trying to quantise general relativity, but they have not succeeded in a fully satisfactory way. Consequently, they have turned to alternative models or extensions, such as string theory and loop quantum gravity.

I'd like to suggest a different strategy. Physicists, limited human beings that we are, construct quantum-mechanical theories by starting with classical theories and quantising them. But presumably nature doesn't work that way; it is simply quantum-mechanical from the start, and classical physics arises as some sort of limiting case. So instead of quantising gravity, perhaps what we should be doing is looking for gravity within quantum mechanics.

What exactly could that mean? Quantum mechanics is a theory of wave functions and entanglement, not of spacetime and particles. We want the latter to *emerge* from the former. We want, in other words, not to put spacetime into our theory as a fundamental ingredient from the start, but to ask how it could be thought of as an approximation to an underlying quantum reality.

To accomplish this ambitious goal, we're certainly allowed to use what we know about the world around us, where spacetime seems to play an important role. Our best current theories, such as the Core Theory, are 'quantum field theories'. They describe fields that fill all of space, so that what you and I would consider 'empty' is really filled with fields that are relatively quiet, in their lowest-energy states. In quantum field theory, even empty space is an interesting place.

In particular, the quantum fields in different parts of empty space are entangled with each other. Regions that are nearby are highly entangled, and regions that are far away are much less entangled (but still a little bit). The notion of 'the distance between two regions' is one that is a defining feature of geometry. In other words, according to the rules of quantum field theory, the amount of entanglement between different parts of empty space is intimately tied to the geometry of that space.

If we want spacetime to emerge from quantum mechanics, rather than being put in from the start, we can imagine turning this around. Our fundamental building blocks are not fields on spacetime, but a wave function of different entangled pieces. Rather than saying 'pieces are more entangled when they are closer together', we can *define* what we mean by 'closer together' as simply 'more entangled'. With that assumption, a curved spacetime manifold can emerge dynamically from a basic quantum wave function. (Note that we're really just thinking about the origin of space, imagining that time is still fundamental. Time might be emergent itself, but that's a subtler and less well-understood issue.)

This idea is still pretty new and untested. But the good news is that it seems to work, at least if we allow ourselves some reasonable assumptions.[7] What we really want is not just to define a spacetime from quantum entanglement; we want to define the kind of dynamical curved spacetime that interacts with matter and energy according to the rules of general relativity. Fortunately, just as we can relate entanglement to geometry, we can also relate it to energy. We can start with the lowest-energy quantum state ('the vacuum') and note that it has a particular entanglement structure. If we modify the entanglement, the energy has to go up (since we're no longer in the lowest-energy state). Therefore, we naturally get a relationship between energy and the curvature of spacetime, just as Einstein suggested many years ago.

There is a long way to go before we can say that this programme of having spacetime emerge from quantum entanglement is the right way to think about quantum gravity. In particular, progress to date hasn't tackled the subtle problems of black holes or cosmology. With that in mind, it's useful to highlight the most important mystery in modern cosmology: the very special conditions that characterised our universe near the Big Bang.

---

[7] C. Cao, S. M. Carroll, and S. Michalakis, 'Space from Hilbert space: Recovering geometry from bulk entanglement', *Physical Review D*, 12 (2016), article 124052.

Mystery 3: The Arrow of Time

Ever since Einstein, physicists have grown accustomed to thinking of time as just part of spacetime. As far as the fundamental laws of nature go, that seems to be correct. Yet we *experience* time and space in very different ways. The biggest difference is that time has a direction; we seem to travel through time from the past to the future, not the other way around. This imbalance between past and future is called the 'arrow of time', pointing towards the future. There is no arrow of space. If you were an astronaut floating freely in your spacesuit, all directions of space would seem equivalent to you. There is no absolute standard by which to judge what is meant by up/down, left/right, forward/backward. But everyone agrees on what you mean by yesterday/tomorrow.

Time's arrow has many seemingly disparate aspects. We all used to be younger, and we grow older, with all that implies. The universe used to be denser and smoother, and it has grown more dilute and lumpier. Life on Earth has evolved from single-celled organisms to a wonderfully diverse ecosystem. We remember the past, which is very different from predicting the future. We think that actions we take today can affect the future, but don't feel the same way about the past. And all these features are seemingly universal. If we someday came into contact with intelligent aliens, they might have different technology or biology or morality, but all of them would remember yesterday and not tomorrow.

The mystery of the arrow of time is *not* where all these arrows come from. We think we know that: they all stem from a single source, the increase of entropy over time.

'Entropy' is a physical concept that has slightly different definitions in different contexts, but for our purposes here, we can think about it as a measure of the disorder or randomness of the arrangement of a physical system. A broken egg has higher entropy than an unbroken one; a well-shuffled deck of cards has higher entropy than one that has been arranged in order. A classic and important example is mixing cream into coffee. When the cream and coffee are separate, they are in an orderly, low-entropy situation; once they are mixed together, the entropy is higher.

You'll notice from these examples that entropy seems to naturally increase over time. Indeed, this is essentially the Second Law of Thermodynamics: in isolation entropy can go up, but it will never spontaneously go down. (The First Law is simply that energy is conserved.) By itself, this isn't mysterious either. Back in the nineteenth century, Austrian physicist Ludwig Boltzmann proposed a useful way to think about entropy: it characterises the number of ways we can rearrange the microscopic constituents of a system while leaving its macroscopic properties unchanged. ('Characterises' here means 'is proportional to the logarithm of', for experts.) When cream and coffee are separate, we can move the molecules around within each one, but not back and forth between the two liquids. As a result, the configuration has low entropy. When they are mixed, however, there are many arrangements of all the molecules in the cream and coffee that basically look the same, so the configuration has high entropy.

From Boltzmann's perspective, it doesn't seem mysterious at all that entropy should tend to increase. There are more ways for a system to be high-entropy than for it to be low-entropy! If you start a system in some low-entropy configuration, evolving towards higher entropy is the most natural thing in the world.

Still, the arrow of time presents very deep mysteries yet to be solved – two of them, in fact. The first mystery stems from the slick explanation we just gave for why it's natural for entropy to increase (because there are more ways to be high-entropy than to be low-entropy). That's true, but there is no direct mention of *time* in that statement. The fact that there are more ways to be high-entropy than to be low-entropy doesn't immediately lead to the idea that entropy increases 'towards the future'. It does seem to imply that, if we have a low-entropy configuration at one moment, then at the next moment the entropy should be higher. But it also implies exactly the same thing about the *previous* moment. Just as there are more ways to evolve towards a high-entropy state, given some low-entropy condition now, there are also more ways to have evolved from a high-entropy state. The fundamental laws of physics are, to the best of our knowledge, completely agnostic about the direction of time. But we think entropy was not higher in the past, we think it's been increasing uniformly for the last 14 billion years, since the Big Bang.

This puzzle was pointed out almost as soon as Boltzmann suggested his theory of entropy, by none other than Josef Loschmidt, who had been Boltzmann's professor at the University of Vienna. It has since become known as Loschmidt's reversibility objection. It is not an unsolvable mystery, and indeed we know a simple way of getting around it: simply posit that entropy started out very low. Conditions near the Big Bang, in other words, were much more orderly than they could have been. This assumption – which we have every reason to believe is correct – has become known as the 'Past Hypothesis'.[8]

At last, we come to the first of our true mysteries associated with the arrow of time: *why* was entropy low near the Big Bang? This is clearly a problem for modern cosmology, one that Boltzmann and his contemporaries had no chance of solving.

Unfortunately, modern cosmologists have not done much better. Many things about the Big Bang remain a mystery to us, and its low-entropy nature is one of the most puzzling features. There are ideas out there, but none of them are complete and compelling. Understanding why the early universe had such low entropy is one of the foremost challenges facing cosmologists today.[9]

Meanwhile, there's a second mystery about the arrow of time that deserves our attention. We've said that all the differences between the past and future ultimately stem from the fact that entropy has been increasing over time. Yet there are many such differences, and they all seem to be quite distinct in some ways. The second mystery is simply this: how does increasing entropy actually account for all the ways in which the arrow of time manifests itself in the universe?

Again, this mystery doesn't seem unsolvable, it's just going to require a lot of work. In many cases we have rough ideas about what is going on, and our task is to turn them into rigorously constructed theories.

Consider, for example, the fact that we remember the past but not the future. What does this have to do with entropy? We're really asking about the general idea that an artefact at the present time – a book, a

---

[8] D. Z. Albert, *Time and Chance* (Cambridge, MA: Harvard University Press, 2000).
[9] Sean M. Carroll, *From Eternity to Here: The Quest for the Ultimate Theory of Time* (New York, NY: Dutton, 2010).

fossil, a photograph – can be correlated with specific information about the past, rather than worrying about the way human memories work. As an example, consider a broken egg that we happen to notice lying on the pavement. As far as we know, there are many possible futures available to that egg. It could just sit there, someone could clean it up, a dog could come by and eat it, and many others. But as far as the past is concerned, we can be more confident about what transpired: there used to be an unbroken egg, and somehow it fell and broke. The broken egg is a simple example of a 'memory'.

Why is there this asymmetry in our conclusions about the past and future? It's not because of the fundamental laws of physics applied to the atoms and molecules in the egg; as far as they are concerned, there are precisely as many possible past histories and future evolutions compatible with our macroscopic information. But secretly, even though we don't usually think in these terms, we also know about the past hypothesis: the universe started with low entropy. So, when we're asking about the past and future of the egg, we're really asking two different questions. The future-directed question is 'given what we see right now, what could potentially happen in the future?' There are many reasonable answers. But the past-directed question is 'given what we see right now, and the fact that the universe began with low entropy, what can we conclude about the past?' That bit of extra information gives us enormous leverage. There are many ways an egg could have ended up on the ground – maybe random atoms and particles in the universe just happened to arrange themselves in an egg-like form. But our assumption that entropy has been increasing all the while pinpoints a specific kind of history: one that involves an unbroken egg not far in the past.

At least, that's the rough idea. Promoting this kind of reasoning to a functioning theory of memories, as well as related phenomena like cause and effect, is an ongoing research problem.

Let's look at one final aspect of the arrow-of-time mystery. As we have seen, the universe starts out with low entropy, and it increases as time passes. And entropy is roughly a measure of disorder. In that case, should we be surprised that there seem to be so many highly organised structures in the universe, like you and me and other biological organisms? If the universe started out smooth and nearly featureless, and all that ever happens is that disorder grows with time, how did all this splendid *complexity* come to be?

This is another case where we have the sketch of an answer, but the details remain to be filled in. The important move is to distinguish between the spectrum of order/disorder and that of simplicity/complexity. These are two different ideas, and they evolve differently over time.

Think back to the earlier example of cream and coffee. We start with a teaspoon of cream, separate from the coffee, then mix them together. At the beginning the system is orderly and low-entropy, at the end it's disorderly and high-entropy. But notice that, at the beginning, it's quite simple – the cream is in the teaspoon, the coffee is in the mug. In the end, it's also quite simple – everything is combined. 'Simplicity' in this context is a way of thinking about how much information you need to completely describe the macroscopic configuration of the system. In either of these cases, it's not much.

In between, however, when you are mixing the cream and coffee with your spoon, things can look quite complex. There are subtle, almost fractal tendrils of cream and coffee mixing. So, the evolution of complexity looks very different from the evolution of entropy. Entropy starts low and increases; complexity starts low, increases, then fades away again.

That's not just a special feature of cream and coffee, it applies to the universe as well. The universe starts near the Big Bang in a low-entropy state, but also a very simple one: hot, smooth, rapidly expanding. Trillions of years from now, after the universe has continued to expand and galaxies have scattered and eventually evaporated away, the universe will be high-entropy, but once again very simple: empty space almost everywhere. It's in between – now – that things are complex and interesting, as the universe is populated by organised structures like galaxies, organisms, and universities.

We shouldn't be surprised that complex structures arise even though entropy is increasing. In fact, the most natural way for such structures to evolve is as an intermediate stage along the path from low entropy to high entropy. But this still leaves important questions. It doesn't seem to be necessary for complexity to arise, it's only possible. What are the general rules according to which complex structures come to be? What kinds of structures naturally evolve? And can this kind of high-level conceptual thinking help us understand the origin of life and consciousness?

Discussion

Most of the effort within modern science is devoted to addressing relatively straightforward, well-defined questions. In the context of fundamental physics, these are things like 'what is the dark matter?' and 'why is there an imbalance between matter and antimatter in the universe?' Not to mention the much larger fraction of physicists who don't work at the 'fundamental' level at all, but focus on questions about atoms or plasmas or materials.

That's as it should be. It makes sense to put a preponderance of one's effort into questions where we have good reason to expect that answers are obtainable, even if it might take a while. Nonetheless, it's also important to keep in mind bigger-picture, conceptual problems, such as the ones I've discussed in this chapter. Human beings have every right to be proud of how much science has discovered about the workings of the universe, but we're clearly not done yet. The revolutions to come are bound to be exciting.

3 Decoding the Heavens: The Antikythera Machine

JO MARCHANT

One of the most exciting and enigmatic objects surviving from the entire ancient world is not a classical statue or vase, or an Egyptian mummy, or even a piece of golden jewellery. It is a collection of battered bronze fragments known as the Antikythera mechanism.

Today, the mechanism is displayed in the National Archaeological Museum of Athens, and the object's history involves a mystery that took over 100 years to decode. The pieces that remain are all we have left of a machine much more sophisticated than anything else known from antiquity; nothing close to its complexity appears before it, or for well over 1,000 years after it. In that sense, the Antikythera mechanism is unique in the historical record, and it baffled scholars for decades after it was discovered. Unravelling how this object was found, how its mystery was, largely, solved, and how answers gleaned from these fragments have shifted perceptions about the capabilities of the ancient Greeks challenges our understanding of the roots of contemporary science and technology.

The modern story of this mechanism started in the year 1900 with a group of sponge divers. The industrial revolution was in full swing, and sponge diving was a thriving industry across the Mediterranean Sea. For thousands of years, sponge divers had swum naked, holding their breath and carrying stones to weigh them down to the seabed. In the 1860s, canvas and rubber diving suits were introduced, complete with bronze helmets. Now divers could breathe compressed air fed through a pipe from the boat above, meaning they could stay on the bottom longer and harvest more sponges. The dangers of breathing compressed air at depth were not yet understood. Over the next few decades, thousands of divers died from the bends, and in many places the suits were banned. But the rewards were great, and fleets of tiny boats full of hopeful, young men still set out every spring, and returned each autumn – those who survived, anyway – with their haul of sponges.

In the spring of 1900, a captain called Dimitrios Kontos and his crew sailed from their home, the island of Symi in the Aegean, towards the coast of North Africa.[1] They spent the summer collecting sponges and sailed home in the autumn. On either the outward or return journey – it is not clear which from surviving historical records – they were blown off course by a storm and sheltered near a tiny island called Antikythera.

Antikythera is a beautiful but barren isle, with rocky, treacherous cliffs. Centuries ago, it swarmed with pirates, but today it is home only to a handful of elderly Greeks. The story goes that, after the storm had eased, one of the divers entered the water but quickly returned, terrified, claiming to have seen a pile of dead women and horses on the seabed. Kontos put on the diving suit (they had just one suit between them) and went down to look for himself. He saw figures on the seabed, about 60 metres down, and realised they were not corpses, but bronze and marble statues. The divers had discovered an ancient wreck, full of treasure. They reported the find to the Greek government in Athens, taking a bronze arm as proof, and were hired to salvage what they could under the direction of official archaeologists. It was a hugely significant project in the history of marine archaeology: the first ever archaeological investigation of an ancient wreck.

The divers worked at Antikythera for 10 months, from November 1900 to September 1901. The weather was terrible, and it was treacherous work; one of the divers died and another was paralysed by the bends. In the end, they returned with the biggest haul of ancient artefacts that had ever been discovered and made news headlines around the world. The finds are now kept in the National Archaeological Museum in Athens. They include bronze and marble statues, gold jewellery, furniture such as ornate bedsteads and pieces of a bronze throne, bronze armour including

---

[1] The story of the sponge divers and their discovery of the Antikythera wreck, and work to salvage its contents, is told in J. Marchant, *Decoding the Heavens: Solving the Mystery of the World's First Computer* (London: Heinemann, 2006), ch. 1; A. Jones, *A Portable Cosmos: Revealing the Antikythera Mechanism, Scientific Wonder of the Ancient World* (New York, NY: Oxford University Press, 2019), ch. 1.

helmets and shields, intricate glass bowls, and heaps of pottery.[2] One of the most impressive finds was a bronze statue of a young man dubbed the 'Antikythera Youth'. It dates to around the fourth century BC and is one of the finest bronze statues known from ancient Greece. There were also huge marble statues which tended to be in worse condition than the bronzes, pitted and eaten away by sea creatures. Some, however, were partly buried in the sediment and these retain beautifully preserved details, such as the crouching limbs and upturned gaze of a young wrestler. One of my favourite finds from the wreck is a bronze head from a statue of a philosopher (identity unknown), dating from the third century BC. It preserves his tousled hair, bushy beard, and piercing eyes inlaid with stone.

The finds were shipped back to the museum in Athens, where staff had the daunting task of sorting through the thousands of pieces. Unidentified or less interesting fragments were left in crates in the museum's open courtyard. It was not until May 1902 that a visitor to the museum noticed a misshapen lump of corroded bronze that had split open.[3] What he saw inside seemed impossible: traces of gear wheels, pointers, measuring scales, and tiny inscriptions in ancient Greek. Photos show the fragments as they now appear, after rounds of cleaning and conservation. The biggest surviving piece, known as Fragment A (Figure 3.1), is particularly striking. It contains a large four-spoked gear wheel with neat triangular teeth cut all around its edge. Several smaller wheels are visible behind. Fragment B (Figure 3.2) contains sections of what look like concentric circular dials. Fragment C (Figure 3.3) reveals two circular scales, precisely marked in a manner similar to a modern protractor, used for measuring angles. Everywhere across the fragments appeared tiny inscriptions in Greek.

It is hard to overstate how incredible this discovery was. No other gear wheels survived from the ancient world, and there was no known

---

[2] The cargo retrieved from the Antikythera shipwreck is catalogued in N. Kaltsas, E. Vlachogianni, and P. Bouyia (eds.), *The Antikythera Shipwreck: The Ship, the Treasures, the Mechanism* (Athens: National Archaeological Museum, 2012), pp. 62–226.

[3] The discovery of the Antikythera mechanism in the Athens museum, and initial studies of it, are described in Marchant, *Decoding the Heavens*, ch. 2 and Jones, *Portable Cosmos*, ch. 1.

FIGURE 3.1 Fragment A of the Antikythera mechanism, held in the National Archaeological Museum in Athens. Photo: J. Marchant.

example of a measuring scale. Yet now, in these pieces, at least 30 wheels were visible, in what was clearly a complex, mathematical device. Nothing comparable to this mechanism emerges in the historical record until the appearance of modern clocks, well over 1,000 years later. Experts were drafted in to study the strange object. They came up with wildly differing dates, from the second century BC to the third century AD, and they failed even to agree on exactly what the device was. Only a few words of the inscriptions were legible, but they included some astronomical terms. Was it an astrolabe? A planetarium? A hoax? During World War II, many of the artefacts from the Athens museum, including the Antikythera mechanism, were hidden under the building's floor to hide them from the invading Nazis. After the war, the frustrating mechanism was never put back on display, and instead sat forgotten in the stores.

FIGURE 3.2 Fragment B of the Antikythera mechanism, held in the National Archaeological Museum in Athens. Photo: J. Marchant.

Scientific Instruments

A man named Derek de Solla Price rediscovered the Antikythera mechanism. Price had been born in east London during the 1920s and trained as a physicist before moving to Yale to become America's first professor in the history of science. He was fascinated by the history of scientific instruments, particularly astronomical devices, and became convinced that understanding the development of such instruments was key to tracing the progress of human knowledge.[4] Price carried out a series of projects, tracing this line of technology back further and further. For one project, he studied a mysterious building in Athens called the Tower of the Winds, built around the first century BC. All that survives of the building today is an octagonal tower, with figures carved on its sides

---

[4] The story of Derek Price's work on the Antikythera mechanism is told in Marchant, *Decoding the Heavens*, pp. 93–158.

FIGURE 3.3 Fragment C of the Antikythera mechanism, held in the National Archaeological Museum in Athens. Photo: J. Marchant.

representing the eight winds. Although the tower is now empty, Price was able to use marks on the marble floor to reconstruct the clock that once stood inside. He concluded that this clock had been driven by water and featured a huge, bronze disc carved with the stars that turned along with the sky.[5] Then Price heard about something far more impressive: the Antikythera mechanism. He travelled to Athens to see it for himself and worked with a Greek epigrapher to decipher more of the inscriptions before concluding that the device was originally contained in a wooden box, with one large bronze dial on the front, and had been turned by a handle on the side. The dial was about 13 centimetres across and around its edge were two concentric scales. On the inner scale Price read 'Chelai',

---

[5] D. J. de Solla Price, 'The Tower of the Winds – piecing together an ancient puzzle', *National Geographic Magazine*, 131 (1967), 586–596; J. V. Noble and D. J. de Solla Price, 'The water clock in the Tower of the Winds', *American Journal of Archaeology*, 72 (1968), 345–355.

the zodiac sign Libra, indicating that this scale showed the 360 degrees of the sky, divided into the 12 signs of the zodiac. The outer scale, on which researchers had previously read the month name 'Pachon', showed all 365 days of the year, divided into 12 months. Having two scales allowed a pointer on this dial to show both the date and the Sun's position against the background stars.[6]

Price also saw individual letters inscribed at various positions around the dial. These referred to further text located around the edge of the dial, which described celestial events that could be observed at the time of year marked by the letter. Price could read snatches of this, such as 'Vega rises in the evening' and 'The Hyades set in the morning.' Similar phrases are known from a type of star calendar common in ancient Greece, called a *parapegma*. The latest reading of the mechanism's inscriptions, published in 2016, shows that at least 42 events were noted here, including the risings and settings of stars and constellations, solstices and equinoxes, and the Sun's entry into different zodiacal signs.[7]

On the back of the device, Price realised there were two dials, one above the other. They seemed to be made up of concentric circles, but he could not figure out what they did, and at this point he was unable to decipher any of the internal workings. Already, however, he grasped the significance of the mechanism. Mechanical gearing was a crucial invention, allowing clocks and accurate timekeeping, as well as the automated machines that drove the industrial revolution. Here was proof that a technology which has shaped the modern world actually had deeper roots in the ancient past. Price wrote in 1957 that the Antikythera mechanism was 'as spectacular as if the opening of Tutankhamun's tomb had revealed the decayed but recognizable parts of an internal combustion engine'.[8] Price struggled to make more progress until the 1970s, when he worked with a radiographer in Athens to X-ray the Antikythera mechanism. The

---

[6] Of course, the solar year is closer to 365-and-a-quarter days long, which is why we have leap years and need to add an extra day every four years. Without this, the months would slowly shift with respect to the seasons. The Antikythera mechanism accounted for this; the calendar scale was rotatable, so that it could be moved around by one day every four years.

[7] Y. Bitsakis and A. Jones, 'The front dial and parapegma inscriptions', *Almagest: The Inscriptions of the Antikythera Mechanism*, 7 (2016), 68–137.

[8] Cited in Marchant, *Decoding the Heavens*, p. 108.

internal structure of the fragments was revealed for the first time in 2,000 years.[9] In the images, Price identified lots more gear wheels and was able to count their teeth. At last, he could start to work out what the device might have been used to calculate.

The mechanism's basic principle is the same as in modern mechanical clocks and watches. Trains of gearwheels that drive each other round can be used to perform mathematical calculations, depending on how many teeth they each have around the edge. To give a very simple example, imagine a gearwheel with 20 teeth, which drives around a wheel with just 10 teeth. For every single turn of the 20-tooth wheel, the 10-tooth wheel will turn twice, meaning you have multiplied your input speed by two (or, to be more precise, minus two, because the speed has been reversed). In a clock, simple trains of gearwheels alter rotation speeds by a factor of 60, to turn seconds into minutes and minutes into hours. In the Antikythera mechanism, the calculations are much more complex. Derek Price worked out the first one by studying a train of six gear wheels leading directly from the input knob. He realised this gear train was taking the input speed, which drives the Sun pointer, and multiplying it by 254/19. Price immediately recognised these numbers from a calendar that Greek astronomers used, called the Metonic cycle. This was a repeating cycle of 235 months, used to harmonise the motions of the Sun and Moon. New moons do not fall on the same date every year, because 12 lunar months do not fit exactly into the 365-day year. But after 235 months, or 19 years, the Sun and Moon finally come back into sync. These are 'synodic' months, which run from full moon to full moon.

Another kind of month, the 'sidereal' month, is equivalent to the time it takes for the Moon to go around Earth one full turn and come back to the same position against the background stars. According to the Metonic cycle, 19 solar years is equivalent to 254 sidereal months. Or, to put it another way, for every 19 turns of the Sun through the sky, the Moon goes through 254 turns. Price realised that the gear train he had measured took the speed of the Sun and converted it into the speed of the

[9] D. J. de Solla Price, 'Gears from the Greeks. The Antikythera mechanism: A calendar computer from ca. 80 BC', *Transactions of the American Philosophical Society*, 64 (1974), 1–70.

Moon. This meant that turning the knob on the side could drive both a Sun pointer and a Moon pointer around the front dial, at their correct respective speeds.

Price died in 1983. His work was continued by a young curator named Michael Wright, who worked at the Science Museum in London. Wright is an amateur mechanic who loves making things, and he is interested in different forms of geared mechanisms.[10] Shortly before Price died, Wright heard about his work on the Antikythera mechanism. The more Wright studied the device, however, the more he felt that Price had not yet found all the answers. There was still more to discover. Because Price's X-ray images had been two-dimensional, this caused difficulties in interpretation. All the layers are squashed together in one image, so the wheels all appear on top of each other, and it becomes hard to tell how they are arranged. Wright travelled to Athens several times in the early 1990s and took his own X-ray images, initially working with a friend, a historian of computing named Allan Bromley. He built a crude tomography machine, a cradle that allowed him to move the object and X-ray film together relative to the X-ray source as he took each image, so that only one plane at a time was in focus.

After Bromley died, Wright worked alone. In his tiny home workshop in London, he pored over hundreds of radiographs and built what he saw. Although it took years, he slowly teased out the arrangement of the wheels inside, revealing features that Price never suspected.[11] For example, Wright carefully measured the back dials and was the first to realise that these were not circular at all, but rather consisted of spiral-shaped slots. Each dial had a pointer with an extendable arm and ended in a drop-down pin, like the stylus on a record player. The pointer started in the middle and, when it reached the outside of the spiral, the user would pick up the pin and put it back at the beginning.

Since the top spiral was divided into 235 sections, Wright realised it must have been used as a calendar, tracking the Sun and Moon across the

---

[10] The story of Michael Wright's work on the Antikythera mechanism is told in Marchant, *Decoding the Heavens*, pp. 159–211.

[11] Michael Wright's research on the Antikythera mechanism during this period is summarised in M. T. Wright, 'The Antikythera mechanism reconsidered', *Interdisciplinary Science Review*, 32 (2007), 27–43.

235-month Metonic cycle that I mentioned above. Inside this spiral are the remains of a much smaller dial, divided into four. At first, Wright assumed that this tracked the longer and more accurate 'Callippic' cycle, which is made up of four Metonic cycles, or 76 years. More recently, however, the inscriptions have revealed a complete surprise. This dial had absolutely nothing to do with astronomy. Instead, it was a four-year dial that showed the timing of different athletics games held across the Greek world, including the Olympics.[12] (A second little dial probably did originally show the Callippic cycle, but it has not survived.[13])

Wright also added further detail to Price's picture of the front dial. Through studying the internal workings of the mechanism, Wright saw that it featured a rotating ball, probably painted black-and-white, that showed the phases of the Moon. Furthermore, when Wright looked closely at the mechanism's big cross-shaped wheel, the one contained in Fragment A, he saw the remains of other structures on its surface. Originally, this large wheel had carried something around. Wright had seen something similar in the astronomical clocks of the Science Museum, incorporated within mechanisms intended to calculate the movements of the planets. The planets move through the sky erratically because they are not travelling around Earth but moving around the Sun at the same time as Earth itself is. Consequently, when the planets are viewed from Earth, they vary in speed and sometimes even stop and change direction. From Greek times until the seventeenth century, Western astronomers explained these erratic motions in terms of 'epicycles', smaller circles superimposed on the larger orbits. Essentially, they thought that planets 'loop the loop' as they travel around Earth. To model this in medieval and renaissance clocks, designers used what is still called 'epicyclic' gearing – small wheels carried around on larger ones.

Had the Greeks thought of this 'epicyclic' gearing first? Wright built a model showing how this might have worked in the Antikythera mechanism. He used trains of wheels riding on top of other wheels, transferring the resulting motion to the appropriate pointer via a slotted lever. As well

---

[12] T. Freeth, A. Jones, J. M. Steele, and Y. Bitsakis, 'Calendars with Olympiad display and eclipse prediction on the Antikythera mechanism', *Nature*, 454 (2008), 614–617.

[13] For example, see Jones, *Portable Cosmos*, ch. 4.

as the Sun and Moon, Wright argued that the mechanism must have had pointers for the five planets known to the Greeks: Mercury, Venus, Mars, Jupiter, and Saturn. If each revolution of the Sun pointer through the zodiac represented a year, then the Moon pointer would have moved around once per month. The planet pointers would each have moved at an appropriate rate with various stoppings and reversals built into the system. Mercury, for example, takes just 88 days to cycle through the zodiac, whereas Saturn takes nearly 30 years. The parts of the mechanism that showed the planets unfortunately do not survive, so Wright was extrapolating from tiny traces. Some experts were initially sceptical that the ancient Greeks could have modelled the planets' movements in such a sophisticated way. Yet inscriptions on the mechanism which have been deciphered in the last few years suggest that Wright was right, since these list the planets and describe their motion, including numbers from precisely the kind of ratios that would have to be involved if Wright's theory was correct.[14]

Technology and Imaging

Wright was deciphering the mechanism's design bit by bit and developing a new view of the device as a planetarium. The bottom spiral dial still baffled him, however, because most of it was missing. Then, around 2004, Wright heard that another team was starting to work on the mechanism, although his rivals could not have taken a more different approach. The Antikythera Mechanism Research Project (AMRP) was an international team, including scholars from a range of fields and using state-of-the-art equipment costing millions of dollars.[15] The team was put together by Tony Freeth, a filmmaker from London. Freeth had heard about the Antikythera mechanism and wanted to make a film about it, but he realised that there was not much new to say (at this point, Wright had not yet published much of his work). Freeth, a mathematician by training,

---

[14] M. T. Wright, 'The Antikythera mechanism: Compound gear-trains for planetary indications', *Almagest: The Inscriptions of the Antikythera Mechanism*, 4 (2013), 4–31. Also discussed in Jones, *Portable Cosmos*, ch. 7.
[15] More information about the Antikythera Mechanism Research Project can be found at http://antikythera-mechanism.gr/.

decided that, if he wanted some new results for his film, he would have to discover them himself. He recruited a team of specialists, including academics from Greece and the UK, as well as researchers from two commercial companies with cutting-edge technology.[16] One of the companies was a small imaging company called X-Tek, based in Hertfordshire, which used high-tech scanning techniques to check components in industry for internal faults. They specialised in a technique called computed tomography, which uses X-rays to scan an object from many different directions, to build a three-dimensional image of its internal structure. To do this for the Antikythera mechanism would require a machine both with very high penetrating power and with very high resolution. At the time, a machine combining the two did not exist, but Freeth convinced X-Tek to build one. They called it BladeRunner because they hoped to use it to image aircraft turbine blades in the future. The resulting machine weighed eight tons, and they shipped it to the Athens museum in November 2005. It was worth the effort. X-Tek's imaging successfully showed the arrangement of gearwheels inside the Antikythera mechanism in breath-taking, three-dimensional detail, down to a resolution of just a fraction of a millimetre.

The other company that Freeth persuaded to participate in his project was Hewlett Packard (HP) in California. HP imaging scientist Tom Malzbender had invented a gadget to model how light bounces off different materials, originally intended for developing computer graphics. Malzbender mounted 50 flashbulbs onto a half-dome, with a camera at the top, pointing down. He placed the dome over an object that he wanted to image and took 50 photographs as the flashbulbs fired in sequence, each one lit from a slightly different angle. Then he fed the images into a computer to build a virtual map of that object's response to light. Malzbender soon realised that, by tweaking the virtual map, he could simulate completely new lighting conditions, scenarios not possible in the

[16] The story of Tony Freeth and the AMRP's research on the Antikythera mechanisms is told in Marchant, *Decoding the Heavens*, pp. 212–260. See also T. Freeth, 'The Antikythera mechanism: I. Challenging the classic research', *Mediterranean Archaeology and Archaeometry*, 2 (2002), 21–35; A. T. Ramsey, 'The latest techniques reveal the earliest technology – a new inspection of the Antikythera mechanism', in *DIR 2007 – International Symposium on Digital Industrial Radiology and Computed Tomography* (Lyon, 2007), www.ndt.net/article/dir2007/papers/2.pdf.

real world. For example, he could make dull stone shine like glossy metal. This technique interested archaeologists, especially those working with Babylonian clay tablets, since it could vastly enhance barely readable markings or inscriptions. After seeing an article on Malzbender's work in *New Scientist* magazine,[17] Freeth convinced him to come to Athens, to try the technique on the Antikythera mechanism.

Malzbender's images highlighted the inscriptions beautifully and allowed far more of them to be read than ever before. (X-Tek's CT images also helped with the readings, by revealing letters that had previously been hidden either beneath the surface of the fragments by the products of corrosion or under different layers of the mechanism squashed together.) The letters on the inscriptions are tiny, some less than a millimetre high, and they are located around the front and back dials, as well as on two separate bronze plates. A 10-year project to read them was published in 2016 and reported 3,400 characters out of an estimated 20,000 that existed originally.[18] One of the team described it as like having a 'whole new manuscript'.[19]

The team confirmed many of Wright's ideas and added more. Crucially, they were able to image a new piece of the mechanism, which had just been discovered in the museum stores. It was part of the mysterious bottom spiral dial. It turned out that this spiral was divided into 223 sections, a number which immediately suggested its function. Greek astronomers used a 223-month period, called the Saros cycle, to predict eclipses: after each cycle of 223 months, or 18 years, the pattern of solar and lunar eclipses roughly repeats. The inscriptions have since confirmed this idea. Little pieces of text within the month segments, called 'glyphs', indicate eclipse possibilities, including whether the possible eclipse is lunar or solar, and the hour of day or night at which it is expected to occur. Like the calendar dial, this spiral too contained a smaller dial, this time divided into three, which represented a longer, but more accurate eclipse prediction cycle called the Exeligmos cycle, which consisted of three Saros cycles, or 54 years. This smaller dial told the user

---

[17] M. Brooks, 'Tricks of the light', *New Scientist*, 170 (2001), 38–40.
[18] See articles in *Almagest: The Inscriptions of the Antikythera Mechanism*, 7 (2016).
[19] From an interview with AMRP astronomer Mike Edmunds in May 2016.

how to adjust the time of day for eclipse predictions in the second and third Saros cycle of each Exeligmos cycle.[20]

As on the front of the mechanism, index letters on the bottom spiral dial directed the user to further information around the edge of both dials. According to the latest interpretation, this text included information about directions (thought to refer either to the direction from which the disc is obscured, or to wind changes during or immediately after the eclipse) and the colour of the eclipse (black or fiery red).[21] This was another intriguing surprise. The accuracy of different astronomical models used in the mechanism varies, but this was the first feature to be discovered that has no factual basis at all: there is no scientific way to predict the colours of eclipses. Directions and colours of eclipses did feature, however, in astrology, not necessarily in individual horoscopes, but in large-scale forecasts concerning the weather, or the fortunes of kings and countries.

The final feature I will mention is possibly the most impressive of all, since it proves that the mechanism also represented the varying speed of the Moon. Because the Moon's orbit around Earth is elliptical rather than circular, the Moon's apparent speed through the sky speeds up and slows down very slightly. The relevant part of the mechanism involves a little component made up of four small wheels, which has been dubbed the 'pin-and-slot'. Two of these wheels are stacked one above the other and are identical, except that they rotate around slightly different centres. The bottom wheel drives the top wheel around via a pin that sticks up into a slot above. As the wheels turn, the pin moves backwards and forwards in the slot – towards and away from the centre of the wheel that it is driving – and this introduces a little 'wobble' into the motion, so that it

---

[20] T. Freeth, Y. Bitsakis, X. Moussas, J. H. Seiradakis, A. Tselikas et al., 'Decoding the ancient Greek astronomical calculator known as the Antikythera mechanism', *Nature*, 444 (2006), 587–591; F. Charette, 'High-tech from ancient Greece', *Nature*, 444 (2006), 551–552; J. Marchant, 'In search of lost time', *Nature*, 444 (2006), 534–538.

[21] M. Anastasiou, Y. Bitsakis, A. Jones, J. M. Steele, and M. Zafeiropoulou, 'The back dial and back plate inscriptions', *Almagest: The Inscriptions of the Antikythera Mechanism*, 7 (2016), 138–215; J. Marchant, 'The world's first computer may have been used to tell fortunes', *Smithsonian Magazine*, 8 June 2016, www .smithsonianmag.com/science-nature/worlds-first-computer-may-have-been-used-tell-fortunes-180959335/.

speeds up and slows down with a set period. It is an elegant feature, and very similar to a mathematical model for the lunar anomaly that was developed by the second-century-BC Greek astronomer Hipparchus.[22] This pin-and-slot component was identified by Michael Wright, but its presence in the device was puzzling. It was sitting on top of a bigger wheel leading to the back of the mechanism that had 223 teeth around the edge, which we now know helped to drive the eclipse prediction dial. Wright realised the pin-and-slot could be used to model the Moon's wobble, but he could not figure out what it was doing on this other wheel.

Freeth had the data from that extra piece of the mechanism and eventually realised the answer to Wright's problem. Freeth worked out that a missing gear had altered the speed of the larger wheel, so that it turned roughly once every nine years. That was the clue he needed; the orientation of the Moon's ellipse is itself rotating around the Earth at exactly this same rate. The Greeks knew that the variation in the Moon's speed shifts through the zodiac in a nine-year cycle. To model this movement in the mechanism, the designer had mounted the pin-and-slot onto the larger turntable. The resulting motion is then fed back to the Moon pointer on the front so that it incorporates a wobble in speed that also moves around the zodiac, just as with the real Moon. After that, another extra wheel re-instated the turntable's original motion so that the eclipse prediction pointer turned at the correct speed. In other words, to reduce the number of gears he needed, the designer had used the turntable twice. The design is extremely clever and beautifully economical. Or, as Freeth put it, 'It's an absolutely unbelievably stunning and sophisticated idea. I don't know how they thought of it. We're just following in the tracks of the ancient Greeks.'[23]

Together, all these researchers and their different approaches have contributed to a modern reconstruction of the Antikythera mechanism which we think includes very nearly everything that was in the original. Following the researchers' stories shows how the identity of the device has changed over time. Price described it as a calendar computer for tracking the Sun and Moon; Wright saw it as a planetarium; for Freeth it was an

---

[22] T. Freeth et al., 'Decoding the ancient Greek astronomical calculator'.
[23] From an interview with AMRP mathematician Tony Freeth, in November 2006.

eclipse predictor. They are all right. I like to think of it a bit like an ancient tablet or iPad, with lots of apps encompassed in one box. One side showed a calendar, the timing of athletics games, and predictions of eclipses, including astrological forecasts. The other displayed the motions of the Sun, Moon, and planets around the sky, as well as the risings and settings of the stars.

Inscriptions recently deciphered by historian Alexander Jones suggest that the planet pointers on the front carried little balls to represent their various celestial bodies: coloured gold for the Sun, for example, or fiery red for Mars.[24] Jones thinks these balls may have been attached at different distances along the pointers, to represent their distance from Earth. If so, this would mean that the dial is itself multifunctional, with a double identity. On the one hand, it showed the changing appearance of the sky as seen from Earth. On the other, it was also a model of the Greek cosmos as seen from outside, from a 'God's-eye view' if you like, with Earth in the middle, the planets moving around it in in their concentric spheres, and beyond them (represented by the zodiac scale) the stars.

A History of Astronomical Models

Half of the mystery, then, appears to be solved: we know what the Antikythera mechanism did, and how it worked. But where did it come from? Who made it? Why? Studies of the shipwreck can provide some clues to answering these questions. The ship was a large trading vessel, perhaps up to 30 or 40 metres long, packed to the brim with a cargo of luxury goods. The bronze and marble statues did not prove helpful in determining the ship's story since they were antiques, some of them centuries old when the ship sailed. Other items were more useful. The cargo included amphoras from Rhodes, Kos, and Asia Minor, and glass-ware from Egypt, perhaps loaded at the trading stop of Rhodes, meaning that the ship came from the east.[25] The most accurate date for the ship's

---

[24] J. Marchant, 'Decoding the Heavens: The Antikythera Machine, the first computer', *Smithsonian Magazine*, February 2015, www.smithsonianmag.com/history/decoding-antikythera-mechanism-first-computer-180953979/. See also Jones, *Portable Cosmos*, ch. 7.

[25] G. D. Weinberg, V. R. Grace, G. R. Edwards, H. S. Robinson, P. Throckmorton et al., 'The Antikythera shipwreck reconsidered', *Transactions of the American Philosophical Society*, 55 (1965), 3–48.

journey comes from the French diving pioneer Jacques Cousteau, who led a team of divers to the wreck in 1976 to make a documentary.[26] They were the first to visit since the sponge divers and had the benefit of scuba gear, which Cousteau had helped to invent. Their aim was to recover any items left buried in the sediment. To excavate a small section of the site, the team used methods including a giant suction hose, literally hoovering up the sea floor and dropping their finds onto the deck of their boat above. They recovered hundreds of small items, including golden earrings, a statuette of a youth on a rotating base, human bones, and bronze and silver coins. The coins were from the cities of Ephesus and Pergamon on the Asia Minor coast, and they dated the shipwreck to around 70–60 BC.[27]

The ship probably departed from Pergamon and then sailed south, probably docking for supplies at the island of Rhodes, which was a common trading stop. Then the vessel headed west but encountered difficulty in a storm and was wrecked at Antikythera. The ship was initially thought to have been Roman. During the first century BC, the Romans were gradually taking over the whole Mediterranean region and carrying shiploads of Greek treasures home to Rome. Very recently, however, a new team, co-directed by archaeologist Brendan Foley, analysed lead components from the ship – such as hull sheathing, sounding weights, and anchors – and found that these originated in either northern Greece or the Cyclades.[28] Perhaps this was in fact a *Greek* trading ship, returning not to Rome but to northern Greece.

Written texts from the period can provide additional clues. There are a few intriguing mentions of devices that sound very like the Antikythera

[26] T. Strauss, *Diving for Roman Plunder*, Cousteau Odyssey 2, VHS (Burbank, CA: Warner Home Video, 1986).
[27] N. Yalouris, 'The shipwreck of Antikythera: New evidence of its dating after supplementary investigation', in J. P. Descœudres (ed.), *Eumousia: Ceramic and Iconographic Studies in Honour of Alexander Cambitoglou* (Sydney: Meditarch, 1990), pp. 135–136.
[28] Personal communication from Brendan Foley, January 2020. For more on the discoveries recently made by Foley and colleagues at Antikythera, see J. Marchant, 'Exploring the *Titanic* of the ancient world', *Smithsonian Magazine*, February 2015, www.smithsonianmag.com/history/exploring-titanic-ancient-world-180953977/; J. Marchant, 'Human skeleton found on famed Antikythera wreck', *Nature*, 537 (2016), 462–463; J. Marchant, 'Antikythera shipwreck yields statue pieces and mystery bronze disc', *Nature*, 4 October 2017, www.nature.com/news/anti kythera-shipwreck-yields-statue-pieces-and-mystery-bronze-disc-1.22735.

mechanism. The most convincing, and earliest, source is Cicero, the Roman lawyer and orator, who wrote extensively on Greek philosophy. He lived in the first century BC and visited Rhodes at exactly the time at which the Antikythera ship sailed. Cicero describes an instrument 'recently constructed by our friend Posidonius, which at each revolution reproduces the same motions of the Sun, the Moon and the five planets that take place in the heavens each day and night'.[29] In another source, he talks about a similar device invented by Archimedes and claims 'it deserved special admiration because he had thought out a way to represent accurately by a single device for turning the globe those various and divergent movements with their different rates of speed'.[30] Until recently, it was hard for historians to know what to make of Cicero's stories. He had no technical training, he did not explain how these devices worked, and some of his descriptions appeared in fictional dialogues. Perhaps he was exaggerating, or fabricated the tales completely. The reconstruction of the Antikythera mechanism shows that a device of just this type existed and suggests that Cicero's accounts were based on reality.

Could the first of Cicero's descriptions reveal the origin of the Antikythera mechanism? Cicero's friend Posidonius was a philosopher who had a school on Rhodes at just this time, when the Antikythera ship would have docked. He was interested in astronomy and could easily have had access to a state-of-the-art technical workshop. Encouragingly, there are other clues that also lead to Rhodes. First, Hipparchus, one of the greatest astronomers of the ancient world, was working on this island just a few decades earlier. Among other achievements, he is thought to have invented the astrolabe, he discovered that the Earth is wobbling slightly on its axis as it spins, and he catalogued the stars. As I noted above, the pin-and-slot component within the Antikythera mechanism models something very similar to Hipparchus' lunar theory. Perhaps it was later made in the same place, and was directly influenced by his work.

Support for an origin on Rhodes also comes from the four-year games dial. Four of the six games mentioned on this dial are major events – Isthmia,

---

[29] Cicero, *De Natura Deorum*, trans. H. Rackham (London: Heinemann, 1933), Book II, ch. 88.
[30] Cicero, *De Re Publica*, trans. C. W. Keyes (London: Heinemann, 1928), Book I, chs. 21–22.

Nemea, Pythia, and Olympia – and would have been known across the Greek world. The other two, deciphered more recently, were much smaller games of only local interest and one of them, Halieia, was held on Rhodes. Jones now suggests that the device was most probably manufactured on Rhodes, perhaps even in Posidonius' workshop. The Halieia mention is like a 'made in Rhodes' label, he says.[31] This is not the whole story, however. The final event mentioned on the games dial, Naa, was held in what is now northwest Greece. The month names that appear on the upper back spiral dial have been shown to derive from the same region. Jones thinks this provides a clue as to why the mechanism was on the ship. Perhaps it was made to order for a customer in northwest Greece, so it had been customised for that region, and was being shipped there but never arrived. This narrative would fit with the latest understanding of the wreck as a Greek trading ship, heading home towards just this region.

The Antikythera device, although it has proved such an enigma to modern observers, is unlikely to represent the beginning of the tradition of geared astronomical models. It was too sophisticated, too confident, too small. Designers surely began with simpler versions, maybe generations earlier, which were then improved over time. This brings us to Cicero's second account, in which he suggests that Archimedes had made such a device. Archimedes lived in Syracuse, Sicily, two centuries before our shipwreck, and he is a good candidate for the inventor of such mechanisms. He was a legendary mathematical and engineering genius, and one of the very few biographical details about him that survives furnishes the detail that his father was an astronomer. One of Archimedes' lost written works was called 'On sphere-making'. The word 'sphere' (*sphaera* in Latin, or *sphaira* in Greek) is thought to have been a generic term used for astronomical models; later authors often call these devices 'Archimedes spheres'.[32]

Researchers are still arguing about whether Archimedes' original model displayed celestial motions on flat dials, as in the Antikythera mechanism, or whether it might have been spherical. In 2015, Michael

---

[31] Jones, *Portable Cosmos*, ch. 4.

[32] For example, see M. Edmunds, 'The Antikythera mechanism and the mechanical universe', *Contemporary Physics*, 55 (2014), 263–285.

Wright displayed a model that he built using the details in Cicero's account, to show that it could have been completed as a sphere.[33] Either way, perhaps Archimedes was the first to come up with the idea of using gears to drive a model of the heavens. Maybe his astronomical devices started off simple and were updated over the generations to incorporate developments in astronomy, such as epicycles and the lunar theory of Hipparchus.

For now, at least, this may be as close as we can get to the people behind the mechanism. It is quite likely that the specific device found in the Antikythera wreck was designed and built by someone whose name is forever lost to history. What was it for? Experts agree that it is hard to see any practical purpose to the device. It does not offer any information useful for navigation which much simpler devices could not have achieved. The front dial, showing the appearance of the sky on different dates, might have been useful for casting astrological horoscopes, but this does not provide any obvious advantage over astronomical tables, and does not explain the mechanism's other functions. One clue is that the inscriptions on the device were clearly designed to be read by non-specialists. They mostly explain what can be seen on the dials; not so much operating instructions but extended captions, as might appear in a textbook or accompanying a museum exhibit.[34] The mechanism seems to have been intended not as a practical tool or a research instrument but as a teaching device; a demonstration of the workings of the heavens for an educated but lay audience. I described it before as like a tablet, with lots of different apps, but it was more than that. It was a portable cosmos, a holistic model that brought together everything that was viewed to be important about the universe at the time: scientific, social, and astrological, in the sky and on the earth.[35]

---

[33] J. Marchant, 'Archimedes' legendary sphere brought to life', *Nature*, 526 (2015), 19; M. Wright, 'Archimedes, astronomy and the planetarium', in C. Rorres (ed.), *Archimedes in the 21st Century: Proceedings of a World Conference at the Courant Institute of Mathematical Sciences* (Cham: Birkhäuser, 2017), pp. 125–141.

[34] Marchant, 'The world's first computer may have been used to tell fortunes'; Jones, *Portable Cosmos*, ch. 6.

[35] For a discussion of the origins and purpose of the Antikythera mechanism, see Marchant, *Decoding the Heavens*, ch. 10. A more recent discussion can be found in Jones, *Portable Cosmos*, ch. 9.

The Roman author Cassiodorus expresses this vision eloquently. In around 506 AD, he wrote a letter to the philosopher Boethius, telling him about an Archimedes sphere. He described it as 'a little machine, pregnant with the cosmos, a portable sky, a compendium of all that is, a mirror of nature ... What a thing it is for a human being to make this thing, which can be a marvel even to understand!'[36] This is one of the last known descriptions of such a device. After that, everything goes quiet. What happened? Where did they go? It seems that, as with so much knowledge from the ancient world, this remarkable technological tradition did not survive the collapse of the Roman Empire. Yet, if we look more closely, there is a sparse trail of clues hinting that at least some parts of the technology did endure. The first is a unique object, in fact the very same object that first got Michael Wright interested in the Antikythera mechanism. It is a sundial from the Byzantine empire with Greek inscriptions, dating from the sixth century AD. It was brought to the Science Museum in pieces by a Lebanese man in 1983, and Wright studied the pieces to build a reconstruction. Portable sundials like this were well known and used the Sun's shadow to tell the time of day. This one has a very important difference: it contained gearwheels. Wright worked out that the sundial's gearwheels drove displays that showed the positions of the Sun and Moon in the zodiac, as well as the day of the month and the phase of the Moon.[37] It was driven by a ratchet that the user turned once each day. The internal workings of this sundial are much simpler than the Antikythera mechanism since it has just eight gear wheels, but the two devices are based on the same principle. The sundial was not a luxury item but an everyday one, suggesting that such devices may have been common across the Greek-speaking world.

A medieval Arabic manuscript that surfaced for sale in London in 2005, before being snapped up by an anonymous collector, provides the next example of anything remotely comparable to the Antikythera mechanism in the historical record. The manuscript is a twelfth-century copy of an original from the tenth century AD and is tentatively attributed to an

---

[36] Cassiodorus, *Variae*, Book I, ch. 45, as cited in Edmunds, 'The Antikythera mechanism'.
[37] J. V. Field and M. T. Wright, 'Gears from the Byzantines: A portable sundial with calendrical gearing', *Annals of Science*, 42 (1985), 87–138.

astronomer called Nastulus, who worked in Baghdad. Part of the manuscript describes an astrolabe, but, once again, this was no ordinary astrolabe, as it featured exactly the same kind of geared calendar as in Michael Wright's sixth-century sundial. Nastulus called it a 'box for the moon'.[38] A physical version of this box for the moon survives as part of a thirteenth-century astrolabe from Isfahan, Iran, now held in the Museum for the History of Science in Oxford. The astrolabe looks normal from the front, but, on the back, there are dials to show the Sun and Moon in the zodiac, the day of the month, and the phases of the Moon.[39] The gearing inside is exactly the same as in the sundial, and in Nastulus's manuscript, yet both astrolabes were made centuries later than the sundial and are inscribed in Arabic, not Greek. This proves that the idea of using gearwheels to model the heavens, which began with the Antikythera mechanism, made it not only into the Byzantine empire, but also into the Islamic world. It is even thought that this technology then travelled back to Europe, where it played a part in triggering the development of modern clocks.

The key creative leap in the invention of clocks was a back-and-forth mechanism called the escapement. This regulates the motion of a falling weight into even chunks, so that it can drive a mechanism at constant speed. The origin of clocks is a puzzle: new technologies usually start simple, then become more complex, but this did not happen with clocks. The first mechanical clocks that appeared in the monasteries and churches of twelfth-century Europe were already huge astronomical display pieces. They were highly complex, with layers of epicyclic gearing and lots of functions. They showed the Sun, Moon, planets, and eclipses, for example, and they told the time almost as an afterthought. Only later did clocks become simplified and streamlined into the timepieces more familiar to us today. The most likely explanation is that, as soon as the escapement was invented, the inclination and knowledge to build astronomical mechanisms was already there, waiting in the wings,

---

[38] Marchant, *Decoding the Heavens*, p. 290.

[39] Astrolabe with geared calendar held at the Museum for the History of Science in Oxford. Muhammad Ibn Abi Bakr, Isfahan, 1221/1222 AD, https://hsm.ox.ac.uk/geared-astrolabe.

ready to be adapted into self-moving models of the heavens. In other words, the Antikythera mechanism is not really a lost technology at all. We can trace a path directly from these ancient gearwheels through to the devices that we still wear around our wrists today.

Specialists are still debating the finer details of the original Antikythera mechanism, such as the precise eclipse prediction scheme it used,[40] or how the designer might have modelled the varying speed of the Sun.[41] Nevertheless, many of the questions surrounding the Antikythera mechanism have now been answered, thanks to the researchers who have devoted large parts of their lives to studying it.

There is one remaining puzzle, though, that is especially intriguing. Which came first: the mathematical theories of the cosmos, or the practical models? Science historians have always assumed that Greek astronomers first came up with mathematical theories – epicycles, for example – and then looked for ways to represent those in bronze. Another possibility, suggested a few years ago, is that in some cases at least, perhaps *the machines inspired the theories*.[42] Maybe a mechanic was playing around with gearwheels, trying to represent the back-and-forth movements of the planets in the sky. He might have realised that a wheel riding on another wheel would do this beautifully. This, in turn, could have inspired the theory of epicycles. In all probability, the two traditions – practical and mathematical – worked together, bouncing off each other. I love the notion that astronomical models, and the mechanics' increasing abilities, were helping to drive and inspire the latest ideas about how the universe works.

This is an insight that goes beyond specific theories or individual celestial motions. Ultimately, I am convinced that models like the Antikythera mechanism helped to inspire and strengthen the overarching idea of the cosmos as a machine, as a physical mechanism that runs

[40] T. Freeth, 'Revising the eclipse prediction scheme in the Antikythera mechanism', *Palgrave Communications*, 5 (2019), 1–12; P. Iversen and A. Jones, 'The back plate inscription and eclipse scheme of the Antikythera Mechanism revisited', *Archive for History of Exact Sciences*, 73 (2019), 469–511.

[41] J. Evans and C. C. Carman, 'Babylonian solar theory on the Antikythera mechanism', *Archive for History of Exact Sciences*, 73 (2019), 619–659.

[42] J. Marchant, 'Mechanical inspiration', *Nature*, 468 (2010), 496–498; Edmunds, 'The Antikythera mechanism'.

according to predictable rules – rules that we can measure, investigate, and understand. That is why this device is so exciting and why it has continued to intrigue scholars and researchers for over a century. Here, in these ancient, battered gearwheels, is an idea that changed humanity. It is a metaphor that became an entire philosophy, that sparked the birth of science and has shaped our modern worldview.

4 Alan Turing and the Enigma Machine

JAMES GRIME

Alan Turing was one of our great mathematicians, a pioneer of computer science and World War II code breaker. This chapter is about him, and the effort to crack the code of the Enigma machines used by Nazi Germany.

Famously, the British code breakers worked in a place called Bletchley Park, halfway between Oxford and Cambridge. During the war it was known as Station X, and it was full of clever people: there were mathematicians, linguists, engineers, and people who just liked puzzles and games because that was the attitude that was needed. By the end of the war, there were 9,000 people working at Bletchley; it was a massive code breaking operation.

There are many code breakers we could discuss, but I'm going to, guiltily, focus on only one: Alan Turing. Turing was one of the lead code breakers and he designed the 'bombe' – a massive machine that would rattle as it deduced the daily Enigma settings. On a good day this would break the code in 20 minutes, so that the code breakers knew what was going on in the German army before the Germans did.

Over the course of this chapter, I will give you a history of Alan Turing, and of the Enigma machine. I will also mention other British code breakers, and Polish code breakers, and describe how the bombe machine worked.

Enigma A

Let's start at the beginning – Alan Turing was born in London in 1912. His father was a civil servant, his mother came from a family of engineers. It was an upper-middle-class upbringing, with his parents spending a lot of time in India, leaving Alan and his older brother in England. Alan loved science as a child, but his interest wasn't necessarily encouraged while he was growing up.

Just a few years after the birth of Turing, the Enigma machine was born as well. In the past, coding had been done with pen and paper, but such codes were either easy to break, or, if made more difficult, they were awkward and complicated to use. At the beginning of the twentieth century, it became possible to mechanise encryption.

The first Enigma-like cipher machine was developed by the Dutch navy in 1915. They decided not to pursue the idea, but it had begun to spread. In 1917, for example, a similar machine was developed by the American Edward Hebern. Unfortunately, it was soon discovered to be breakable.

In 1918, a German engineer called Arthur Scherbius patented his own rotor cipher machine: the Enigma machine. The German navy wasn't interested in it, so Scherbius sold the idea to a security firm, which set up the *Chiffriermaschinen Aktien-Gesellschaft* (Cipher Machines Stock Corporation), with Arthur on the board of directors. In 1924, the company produced its first machine, 'Enigma A'.

Enigma A worked like a typewriter: it could code and decode, and it had a 'neutral' setting, which meant it could actually be used as a typewriter. This was office equipment, sold to people who needed to keep secrets, like banks and railway companies. But the machines were expensive and didn't sell well. Then, in 1925, the German navy began using them.[1]

Meanwhile, in 1926, Turing started going to Sherborne School in Dorset. The first day of term coincided with the 1926 General Strike in Britain, but so determined was he to attend his first day that he rode his bicycle unaccompanied for more than 60 miles, from Southampton to Sherborne, stopping overnight at a pub. The story got into the local papers.

---

[1] According to some accounts, around this time, a lot of the British Generals from World War I were publishing their memoirs, with some revealing that the United Kingdom had broken German ciphers in World War I. It was then that the German army realised they needed something better and started using the Enigma machine. One of the people who did this in his memoirs was Winston Churchill – so he only had himself to blame!

Unfortunately, Alan didn't have a particularly easy time at school, being far more interested in science than in classics and arts. This is clear from comments in Turing's school reports,[2] such as this one from his House Master:

> He has as good brains as any boy that's been here, and they are good enough for him to get through even in 'useless' subjects like Latin and French and English. His manner of presenting work is still disgusting and takes away much of the pleasure it should give. He doesn't understand what bad manners bad writing and messy figures are.

His maths report wasn't much better:

> Not very good. He spends a good deal of time apparently in investigations in advanced mathematics to the neglect of his elementary work. A sound ground work is essential in any subject … His work is dirty.

And, from his headmaster:

> I hope he will not fall between two stools. If he is to stay at a Public School he must aim at becoming educated. If he is to be solely a scientific specialist, he is wasting time at a Public School.

Despite all this, Turing went to study Mathematics at King's College, Cambridge in 1931.

Enigma I

Meanwhile, during the time Turing was at school, the German army had joined the navy in using the Enigma machine. In 1929, the inventor of Enigma, Arthur Scherbius, died when his horses bolted while he was driving a carriage. The company continued without him, and, in 1930, the German military began using a new version of Enigma, called 'Enigma I'. This was a model that was just for the military.

[2] A. Hodges, *Alan Turing: The Enigma* (Princeton, NJ: Princeton University Press, 2012).

Enigma I looks and works very much like a typewriter, except that there are two sets of letters. One is the keyboard, and above that, instead of paper, is a second set of letters called the lampboard. When someone types a message on Enigma, the code letters light up on the lampboard. For example, if I want to send the word HELLO, then I type HELLO on the keyboard and the following letters might light up:

HELLO
GFPDS

Note that the machine itself does not transmit; the operator has to write the code letters on a piece of paper and then transmit them by radio.

You may have noticed something strange about the above: the two Ls in HELLO became two different letters. That's because each letter of the message is made with a different code. For example, if I press the letter A repeatedly, it will be different each time. There's no pattern to this and there's no way to know what comes next. That is why the code was thought to be unbreakable. Let's have a look at what's going on inside the machine.

Inside the Enigma machine, there are three wheels, called rotors (see Figure 4.1). Inside each rotor, there is lots of criss-cross wiring, all mixed up like spaghetti inside. Figure 4.2 shows what Enigma wiring looks like, using just the letters 'a' to 'f'. If you follow the path from 'a' through the machine, you will go through the plugboard, the three rotors, then hit the reflector, which sends you back through the machine in reverse to finally arrive at 'e'. So 'a' becomes 'e' in the code, and equally 'e' becomes 'a'.

It should be noted that, in the real Enigma machine, each rotor has 26 starting positions, and the three rotors can come out and be swapped over. A plugboard sits at the front of the machine, which looks like an old-fashioned telephone switchboard. This military addition added an extra level of scrambling by connecting 12 letters into six pairs using wires.

All this increases the complexity, but the most important feature is that the rotors move when a message is typed. The rotor on the right moves one place forward every time you press a key. When the right-hand rotor has done a full revolution, it will kick the middle rotor one place forward. When the middle rotor does a full revolution, it will kick the left-hand

(a)

FIGURE 4.1 (a) Three-rotor army Enigma machine, and (b) close up of the
rotors. Photos: The Enigma Project.

rotor one place forward. In other words, there's a fast rotor, a middle rotor,
and a slow rotor.

Enigma is a very clever machine, but it's essentially just a big circuit.
There is a battery inside and when a letter is pressed, this battery
connects to a light – so it lights up. But the wires are inside the rotors,
so, when I let go, it disconnects the battery, and the wires rotate; and
when I reconnect the battery, it connects to a different bulb. That's why
the code changes for each letter of the message.

(b)

FIGURE 4.1 (*cont.*)

However, Enigma is no good if we can't decode a message. So, let's first send a short message, like 'HI'. Type HI on the Enigma machine and you get the code, for example WP. Transmit the code by radio, and miles away another German officer is listening to the signal and writes down the code. They will have a second Enigma machine, exactly like the first one – that is, with rotors and plugboard in the same setting. If the setting is indeed the same, when I type the code WP, I get the original message HI back.

Effectively, Enigma always turns the 26 letters of the alphabet into 13 pairs, which means encoding something twice results in the original

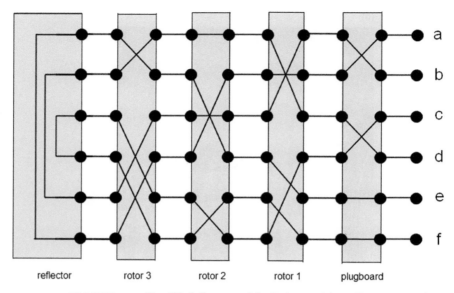

FIGURE 4.2 Simplified diagram of the Enigma wiring. Illustration and copyright: The Enigma Project.

message, making coding and decoding that much easier. Hence, Enigma is both a coding machine and a decoding machine: type in a message and get the code, type in the code and get the message back.

This works only if the two machines are set the same, so each operator had the settings written down on a piece of paper, called a keysheet (Figure 4.3), which indicated how to set the machine that day. Let's go through it:

1. The first column is the date.
2. The second column is the order of the rotors. Remember that there are three rotors, and they can be swapped over. There are six ways to arrange three rotors.
3. The third column is called the ring setting. On the outside of each rotor is a Bakelite ring that labels the rotor position from A to Z. However, these rings can move. This doesn't change the code in any way, but it changes the labelling for that position. There are 17,576 ways to set the three rings.
4. The fourth column is the plugboard, which connects 12 letters into six pairs. There are 100 billion ways to do that.

Geheim! Sonder – Maschinenschlüssel BGT

Datum	Walzenlage	Ringstellung	Steckerverbindungen	Grundstellung
31.	III II I	F T R	HR AT IW SK UY DF	vyj
30.	III I II	Y V P	OR KI JV OE ZK KU	oqr
29.	II III I	O H R	UX JC PL LK TA ED	vhf

FIGURE 4.3 Example of a keysheet. Photo: The Enigma Project.

5. The last column is the ground setting, which is essentially the rotor starting position. Again, there are 17,576 ways to do that.

The ring setting (number 3 above) does not contribute to the total number of settings; it just makes the set-up a little more complicated. Ignoring that, altogether there are 10,000 trillion ways to set Enigma. For any code breakers, this would be too many to check.

However, it is not quite so simple, because there would be hundreds of Enigma messages every day. If all those messages used the same setting, it would become easier to break. Instead, each operator was allowed to pick their own rotor starting position, which could be different for each message. But that meant the operator had to put the secret starting position at the beginning of the message – in code. And they would do that using the Enigma machine itself.

First, the operator would set their machine to the ground setting indicated on the keysheet. For example, let's say the ground setting is XYZ, so the operator sets the positions of the rotors to XYZ. Then, the operator picks their own secret setting, say ABC. They would type ABC into the Enigma machine to make a code. In fact, they would type the setting twice, so ABCABC might become JTEQGL. These six letters were then sent at the beginning of the message.

At the other end, the second operator would set their machine to the ground setting, XYZ, and decode the first six letters to get ABCABC. Now the operator knows that ABC is the secret setting to decode the rest of the message.

This is certainly a complicated sounding procedure, but necessary to keep messages secure.

Code Breaking in Poland

By the early 1930s, Poland knew Germany was a threat and had begun trying to break the Enigma. Before this time, code breakers tended to be linguists, as code breaking requires certain language skills. Enigma, however, was a technological problem, and needed a technological solution.

In 1932, the Polish Cipher Bureau took on several young mathematicians, including Marian Rejewski, Henryk Zygalski and Jerzy Różycki. The code breakers were provided with a commercial Enigma machine, so they were familiar with the mechanism. Unfortunately, however, the military machine was wired differently from the commercial one, and it included the new plugboard. To help them get started, the code breakers had several messages in both Enigma code and original plaintext. Some of these messages had been collected by spies, and one example came from an Enigma sales brochure.

Using these decrypted messages, Rejewski managed to set up some equations to work out the wiring of the military Enigma. Unfortunately, equations were not enough: the Polish code breakers still needed data to feed into the equations, preferably some German Enigma keysheets. To get these, they needed help from a spy.

Hans Thilo-Schmidt was from an aristocratic family but had fallen on hard times. His brother was a general in the German army and helped him get a job: Hans had to destroy top secret Enigma keysheets. Instead of doing that, he ended up selling them to the French, who passed them on to the Poles.

This was exactly what Rejewski needed. Using the keysheets, he was able to deduce the rotor wiring for the military Enigma machine!

There was one cause of concern. After deducing the wiring for the Enigma rotors, the code breakers put the new rotors into the machine and . . . it didn't work. This was because, on the commercial machine, the rows of letters on the keyboard were connected to the first rotor in alphabetical order. That meant Q became A, W became B, E became C, and so on. But now the Polish code breakers had discovered that the

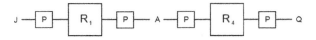

FIGURE 4.4 Example wiring of two Enigma machines. P denotes the plugboard, and R1 the rotors, set to position 1. Illustration and copyright: The Enigma Project.

military machine was different! The keyboard was connected to the first rotor in some different order, and the number of ways to do that is a whopping 400 million billion billion. Then Rejewski thought, 'Have they been so stupid as to connect A to A, B to B', and so on. And they had! By 1933 the Poles had a working Enigma replica, which they called an 'Enigma double'.

However, deducing the machine was only half the problem. The other half was working out the settings, which changed every day. Let's describe an earlier method to work these out.

We saw before that at the beginning of each message there is a secret setting, for example ABCABC becomes JTEQGL. In other words, starting from the beginning, pressing J on the machine gives A, and then, three steps later, pressing A on the machine gives Q.

This means we can wire up two Enigma machines, as shown in Figure 4.4. Figure 4.4 is just treating Enigma as a magic box – in fact I've split up one machine into three boxes. First the letter 'J' goes through the plugboard (labelled P), then the result travels through all the rotors, set in position 1 (labelled R_1), and finally the result goes through the plugboard again, which produces the letter 'A'. The second machine is similar, but it has the rotors in position 4 (R_4).

Summarising, A becomes J with the rotors in the first position, and A becomes Q with the rotors in the fourth position. Altogether, J becomes Q in this set-up. That's a clue: we can't predict what one Enigma machine will do, but we can deduce what two Enigma machines will do when hooked together.

A different message might have a secret setting QBMTOZ, which tells us Q becomes T; a third message might have the setting TPNJQC, which tells us T becomes J. Together, these three secret settings make a cycle (JQT).

Now, you may be receiving dozens of messages every day. Keep going and you'll get cycles for the whole alphabet. For example, (APCFKYLXS)

(DUHEZMGNV)(BOR)(JQT)(I)(W). The lengths of these cycles were called the 'characteristic'. The characteristic of this example is 9, 9, 3, 3, 1, 1.

Each setting has its own set of cycles. This is like a fingerprint, although not quite as unique. If you can make a catalogue of these cycles for each Enigma setting, then you can look up which setting is being used that day.

As described so far, the card catalogue would still need to list 10,000 trillion Enigma settings. Fortunately, the plugboard does not affect the characteristic and can be ignored. This means the Polish code breakers only needed to catalogue the characteristic for each rotor position and rotor order, which is about 105,000 characteristics, relatively a much smaller number. Any other settings are then found by hand.

It took the Poles a year to catalogue all the settings, which they finished in 1937. By 1938 the Polish cipher bureau was reading about 75 per cent of Enigma messages. Then the Germans changed their procedures.

In 1938, the Germans stopped using a daily ground setting to encrypt the secret message setting. Instead, each operator was allowed to pick his own ground setting, and to send that in plain at the beginning of the message, before sending the encrypted message setting as before. So, at the beginning of each message were nine letters that might look like this: GKD WAVWHA. Here GKD is a ground setting chosen by the operator and sent in plain, and WAVWHA is the coded message setting.

This change meant that the catalogue of characteristics was now useless: new ideas were needed. This battle between German code makers and Polish code breakers continued throughout the 1930s, until Rejewski designed the 'bomba'.

Some say the bomba was named after a type of ice cream, some say it was named because of the ticking noise it made, and others say it was simply as good a name as any. The bomba was small enough to stand on a desk, and worked like six simultaneous Enigma machines, cracking the secret six-letter message settings at the beginning of each transmission.

The code breakers would search these coded six-letter settings and look for any that had the same letter in the first and fourth positions. For example, earlier we saw the message setting GKD WAVWHA, which has W in the first and fourth positions. They would then look at other messages to find any examples with the same letter, W, appearing in the second and fifth positions, and the third and sixth positions.

This means that, if we set up six Enigma machines in the correct positions, they will all output W. And that is how the bomba worked. It was six Enigma machines, without the plugboards, checking all 17,576 rotor positions. When they all have the same output, we know that we have found the correct rotor settings.

However, a bomba machine was needed for each rotor order. So, with three rotors, you needed six bomba machines. At the beginning of 1939, the German army added two more rotors, meaning the operator would now choose three from a box of five. That meant that the number of rotor orders went from 6 to 60, so they would need 60 bomba machines. Meanwhile, the number of wires on the plugboard also went up from 6 to 10. Together, these changes increased the number of settings by 10,000.

This stretched the Polish resources to the limit: five weeks before Poland was invaded, there was a secret meeting with the British code breakers, and the Poles passed on their work and a couple of Enigma replicas.

The Naval Enigma Code

Meanwhile, as the Polish code breakers battled Enigma in the 1930s, Turing was studying maths at King's College, Cambridge. He graduated with a first-class degree in 1934 and was elected fellow of the college on the strength of his dissertation, in which he provided a proof of the Central Limit Theorem (an important result in statistics), not knowing that this proof had already been discovered in 1922. It was around this time that Turing decided to tackle one of the big unsolved problems of the twentieth century, known as the Decision Problem.

At the beginning of the twentieth century, the German mathematician David Hilbert set out a number of challenges to mathematicians. One of these problems is known as the Decision Problem (*Entscheidungsproblem*). In broad terms, it goes like this: is there a single mathematical algorithm that can solve any mathematical question? In other words, if you have a mathematical question, apply this algorithm and eventually it will tell you 'true' or 'false', or 'provable' or 'not provable'.

To solve this problem, Turing conceived of a hypothetical machine that could do any calculation a person can do. Given a problem, the machine will either stop and give you an answer, or it will run forever if an answer doesn't exist. If we can determine whether the machine will stop, that will tell us whether the problem is solvable. However, Turing was able to show that there is no general method for determining whether the machine will stop. This is a definitive example of an 'undecidable problem'.

With this work, Turing had solved one of the big problems of the twentieth century: there isn't a general algorithm that can solve all mathematical questions. As an unintended consequence, he defined the mathematics of modern computing. Turing was 22 at the time.

After this triumph, Turing went to Princeton for his PhD. When he returned two years later, he started working part-time for the British code breakers. The year was now 1939, and war with Germany seemed inevitable.

Indeed, the day after war was declared, Turing reported for duty at Bletchley Park. He was put in charge of Hut 8, a team responsible for breaking the naval Enigma code.

As noted, each Enigma message started with a secret rotor setting written in code. To do this, the German army and air force would use the Enigma machine itself, repeating the setting to make six letters of code. The navy did things differently. Like the army and air force, they would start with a secret setting, but instead of using Enigma, they used a separate code book.

The Polish code breakers had tried to figure out the navy code book, but they were only partially successful. As the war started, this was still an open problem. Indeed, working out the naval procedure was one of Turing's first contributions to breaking Enigma. Let's describe the procedure first.

From a book, the operator picks two triplets of letters. For example, HLG and KQK, called the 'key indicator group' and 'message indicator group':

	H	L	G	
K	Q	K		

Now the operator would fill in the spaces with two dummy letters of their choosing:

A	H	L	G
K	Q	K	Z

This makes four pairs: AK, HQ, LK, GZ.

A second code book, known as a 'bigram table' would encode the pairs into different pairs. For example, using the bigram table, AK might become BD, HQ becomes BJ, LK becomes EM, and GZ becomes EJ. The operator would then send the result, BDBJ EMEJ, at the beginning of the message.

At the other end, the receiver decodes these eight letters with their bigram table. The key indicator, HLG, confirms both users are using the same daily setting, while the message indicator, KQK, is typed into the Enigma machine on the daily ground setting. For example, typing KQK into Enigma might result in IYS.

The output, IYS, generated by the Enigma machine, is a lot more random than any triplet of letters chosen by the operator, and so it is this result that was used as the rotor setting for the rest of the message.

Turing had deduced much of this procedure himself, as well as partially reconstructing the naval bigram tables. A later pinch of Enigma documents from a German ship confirmed Turing was right.

However, to reliably break Enigma, the code breakers still needed to get hold of the complete bigram tables. Part of the problem with capturing these tables was that they were written on blotting paper, which meant that the book would dissolve if a ship was torpedoed, and the book got wet.

A plan was drawn up: 'Operation Ruthless', devised by Ian Fleming, personal assistant to the Director of Naval Intelligence and later author of the James Bond novels! In a 1941 document,[3] Fleming outlined the plan as follows:

> I suggest we obtain the loot by the following means:
>
> 1. Obtain from the Air Ministry an air-worthy German bomber (they have some).

[3] Reproduced in full in S. McKay, *Bletchley Park: The Secret Archives* (London: Aurum Press, 2016).

2. Pick a tough crew of five, including a pilot, W/T Operator and word perfect German speaker. Dress them in German Air Force uniform, add blood and bandages to suit.

3. Crash plane in Channel after making S.O.S. to rescue service in P/L.

4. Once aboard rescue boat, shoot German crew, dump overboard, bring rescue boat to English port.

In order to increase the chances of capturing an R. or M. with it's [sic] richer booty, the crash might be staged mid-channel. The Germans would presumably employ one of this type for the longer and more hazardous journey.

And they nearly did it! They got the bomber aeroplane, Fleming took his team to Dover and waited. What they were waiting for was for the right type of ship to pass by, but no suitable naval vessel could be found, and the operation was called off.

It's a fascinating story of espionage, and clearly the inspiration for the James Bond books – especially *From Russia with Love* in which Bond has to obtain a Russian code machine named Spektor.[4] All this is a thinly veiled reference to Enigma.

Code breaker Frank Birch later wrote about the disappointment Turing, and another Bletchley code breaker Peter Twinn, felt about this cancellation:

> Turing and Twinn came to me like undertakers cheated of a nice corpse two days ago, all in a stew about the cancellation of operation Ruthless. The burden of their song was the importance of a pinch.[5]

Fortunately, other missions did result in capturing the information Bletchley needed.

Turing and the Bombe Machine

At Bletchley Park, Turing was known as 'The Prof', despite only being 28 at the time. He had a reputation as a bit of an eccentric. Famous stories

---

[4] The machine is called Lektor in the film version as Spectre has become the name of Blofeld's secret organisation.

[5] Mavis Batey, *From Bletchley with Love* (Bletchley Park: Bletchley Park Trust, 2008), pp. 2–6.

of Turing at Bletchley include him cycling into Bletchley wearing a gas mask to avoid hay fever and chaining his mug to the radiator pipes to prevent it being stolen. Both stories are usually told as examples of how eccentric he was – but they sound perfectly reasonable to me!

Turing was also a talented long-distance runner, and he would occasionally run for meetings from Bletchley to London, which is 40 miles. This was not some casual hobby, and only an injury prevented his serious consideration for the British team in the 1948 Olympic Games. I think it is revealing of his single-mindedness.

During 1941, code breaking at Bletchley Park was hindered by staff shortages. So, going over the heads of those in command at Station X, Turing and other leading code breakers at Bletchley wrote a letter directly to Winston Churchill. To his credit, Churchill wrote to his Chief of Staff:[6]

> ACTION THIS DAY: Make sure they have all they want on extreme priority and report to me that this had been done.

This, I believe, shows that Winston Churchill completely understood the importance of the work done at Bletchley Park, when other world leaders might not have. It was this personal support that allowed Bletchley to do what they did.

Also in 1941, Turing asked one of the women in Hut 8 to marry him. This was Joan Clark, who was a talented mathematician herself, having been recruited from Oxford University. The engagement was short-lived after Turing admitted he was gay. Joan was reportedly unfazed by the revelation. Since homosexuality was still illegal, Turing had learnt to be discreet, and to reveal his sexuality only to a select few people. Turing's secret was made public in 1952, after his house was burgled and Turing naively admitted he suspected a former lover.

After his arrest for gross indecency, Turing was treated very badly. He was considered a security risk and lost his job consulting for the GCHQ code breakers. He was barred from the United States and offered the choice of prison or hormone treatment (he chose the latter).

---

[6] Reproduced in full in McKay, *Bletchley Park.*

According to the official report, Turing committed suicide by lacing an apple with cyanide and taking a bite. His mother never thought it was suicide, instead believing his death to be accident caused by experiments with cyanide at home, and it is true that the apple found by Turing's bed was never tested.

But let's return to Turing's work at Bletchley Park, and his most important contribution to the daily breaking of Enigma: the design of the bombe machine.

Up until now, the Polish and British code breakers were breaking army and air force codes using the secret starting position at the beginning of every message. If the military changed their procedures, this method would no longer work. A more robust strategy was needed to break the codes.

The new method exploited an inherent flaw in the machine. As stated before, the Enigma machine turns the 26 letters of the alphabet into 13 pairs, which means that no letter can be paired with itself. For example, if the operator pressed the letter 'a' repeatedly, every code letter would eventually light up, except 'a' itself. And that was a clue.

There were listening stations around the world that tuned into the German frequencies, writing down the coded messages, which were consequently delivered to Bletchley Park. The code breakers would then try to guess a word or a phrase that might be in one of those messages.

For example, at 6 am every morning, the Germans would transmit a weather report. This message was written in a standard format, so it would be almost the same every day. Let's guess a phrase from that message and try to find where it fits in the code.

We simply use the phrase 'weather report', or *Wetterbericht* in German, and slide that phrase underneath the code, from left to right, trying to find where it might fit – but remember, a letter cannot become itself!

			1	2	3	4	5	6	7	8	9	10	11	12	13			
...	j	x	a	t	q	b	g	g	y	w	c	r	y	b	g	d	t	...
			w	e	t	t	e	r	b	e	r	i	c	h	t			

These phrases were known as 'cribs'. As you can see, if we move the crib one place to the left, the first 't' will match the 't' in the code – which is not possible. On the other hand, if we move the crib one place to the right, the second 'r' will match with 'r' in the code, which is also impossible. The current position might work though; it's still a guess, but certainly there are no matching letters.

The code breaker would then draw a diagram describing the relations between letters, called a 'menu' (Figure 4.5). The menu is the same information as the crib, although menus can contain cycles and are more useful.

We're now going to use this to deduce the plugboard at the front of the machine. As you can see from the crib and the menu, in position 2, 't' becomes 'e'. A simplified drawing of an Enigma machine, as shown in Figure 4.6(a) (the notation is the same as in Figure 4.4), represents what is happening.

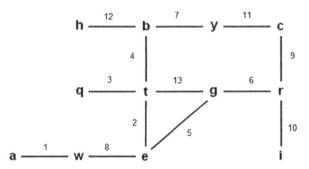

FIGURE 4.5 Diagram describing the relations between letters on an Enigma machine, called a 'menu'. Illustration and copyright: The Enigma Project.

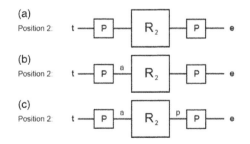

FIGURE 4.6 Three simplified drawings of an Enigma machine to help deduce the plugboard settings. Illustration and copyright: The Enigma Project.

Now we are going to make another guess (Figure 4.6(b)). This time we will guess one of the wires on the plugboard. These wires connect one letter to another to make a pair. For example, let's assume (ta) is a pair on the plugboard. Then, after passing through the plugboard, input 't' becomes 'a'.

Now the code breakers know how the rotors are wired. So, they just need to pick a position for the rotors and send the letter 'a' through and see what happens. Let's say 'a' becomes 'p' (Figure 4.6(c)). This helps us deduce another plugboard wire because, if the final output is 'e', then (pe) must be another plugboard setting.

In the same way, by considering what happens in the 3rd, 4th, and 13th positions, we get three further plugboard settings, namely (kq), (xb), and (tg) (Figure 4.7).

But there's a problem here. We have deduced that (ta) and (tg) are both connected on the plugboard. Yet, 't' cannot be connected to two different letters at the same time, so something has gone wrong. What has gone wrong? It was our assumption at the beginning. We assumed (ta) was a plugboard setting. But it can't be, so we guessed wrong.

If 't' isn't connected to 'a' on the plugboard, then maybe it's connected to 'b', or 'c' or 'z' – in fact there are 26 possibilities to check (including 't' connects to 't', which means the letter is unplugged). If none of these

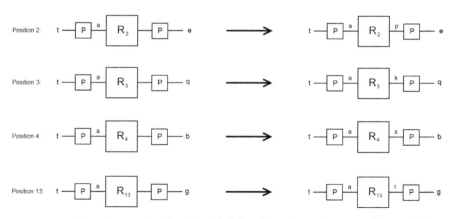

FIGURE 4.7 Further simplified drawings of an Enigma machine to help deduce the plugboard settings. Illustration and copyright: The Enigma Project.

possibilities work, then it is our rotor position that's wrong, and we should check the next one.

That sounds very laborious, but there are a couple of ways to speed up the process. First, once you've found a contradiction, all other consequences are equally false. Hence, the other plugboard settings we found, like (pe), (kq), (xb), and (tg) are also false and don't need to be checked. Once we find a contradiction, we want to find as many consequences of that as possible so they can all be rejected.

Secondly, these deductions can be made instantaneously by electrical circuits. For example, if we apply a current to our previous guess, (ta), then current will flow instantaneously through (tg) and all the other plugboard settings we found.

Turing used this principle in his bombe machine, which acted as simultaneous Enigma machines connected in series. Logical implications from one 'Enigma machine' would be fed into the other machines, set at different positions determined by the menu. If the current flows through only one plug for a given letter, or all but one plug, then the machine would stop, as that might be a valid position. This position was then noted, and checked on an Enigma machine, or Enigma replica.

The bombe machine could check the logical implications of all 17,576 rotor positions in 20 minutes. The machines themselves were designed with the help of fellow code breaker Gordon Welchman, and Doc Keen, chief engineer of the British Tabulating Machine Company.

The first bombe was installed in March 1940 and was called 'Victory'. Two months later, the German procedures changed. Message settings were no longer to be repeated, meaning all other methods used before the bombe no longer worked. The bombe proved to be a successful attack on Enigma for the remainder of the war.

Let me finish with a few things you may have heard about. At the beginning of 1942, the German U-boats introduced a four-rotor Enigma machine. After the British were so successful at sinking supply ships, a suspicious Admiral Dönitz, commander-in-chief of the German navy, asked for an increase in security. This change was a problem for the code breakers of Bletchley Park, as it locked them out of the code for 10 months, when the allies crucially needed ships from the USA to cross the Atlantic.

Eventually, it was Turing who deduced the wiring of the fourth rotor, as well as the fact that the fourth rotor effectively had a neutral position, so it could work with three-rotor machines. This allowed the code breakers to break messages from the U-boats to stations that had a three-rotor machine as normal, then deduce the position of the fourth rotor by hand. Meanwhile, cooperation between Bletchley Park and the Americans led to the USA constructing four-wheel bombe machines specifically to break the U-boat messages.

Enigma was used throughout the German military. There was, however, another code machine used by Hitler and the top level of the Nazi party that was more difficult than Enigma. It's true that Turing did some work to deduce the settings of this machine, called Lorenz, but the essential work was done by a young mathematician called Bill Tutte. He devised a way to break Lorenz, although the method was far too difficult to do by hand. To solve this problem, Post Office engineer Tommy Flowers designed and built Colossus, arguably the world's first computer. It allowed Bletchley Park to read messages from Hitler himself.

Meanwhile, the Polish code breaker Marian Rejewski had escaped Poland and was still breaking codes for the Polish government in France. In 1943, he escaped to Britain, where he continued his work. Bizarrely, despite requests from Bletchley, the British and Polish code breakers were not allowed to talk to each other.

As for the Enigma machines themselves, after the war the allies started to collect them and sell them to foreign embassies, providing a machine that they could break. Other machines continued to be used up to the 1950s, but eventually most were lost or destroyed.

Turing himself ended the war acting more as a consultant. He went to America to liaise with the American code breakers who were now building their own bombe machines. When he returned, he started to develop a speech enciphering machine called Delilah, which is certainly a forerunner to the technology we use today.

In 1948, Turing went to Manchester University, where he worked on computing, artificial intelligence, and mathematical biology. He received an OBE for services to the Foreign Office, but the true reasons for the honour remained secret for many years.

During the early years, Turing's second-in-command was Hugh Alexander, a mathematician who later took over the day-to-day running of Hut 8. Alexander has been quoted as saying:

> There should be no question in anyone's mind that Turing's work was the biggest factor in Hut 8's success.
>
> In the early days he was the only cryptographer who thought the problem worth tackling and not only was he primarily responsible for the main theoretical work within the Hut but he also shared with Welchman and Keen the chief credit for the invention of the Bombe.
>
> It is always difficult to say that anyone is absolutely indispensable but if anyone was indispensable to Hut 8 it was Turing.
>
> The pioneer's work always tends to be forgotten when experience and routine later make everything seem easy and many of us in Hut 8 felt that the magnitude of Turing's contribution was never fully realized by the outside world.[7]

After the war, Winston Churchill spoke more generally about the British code breakers, describing them as 'the geese that laid the golden eggs, but never cackled'.

Certainly, their work was extremely valuable, but top secret. After the war, Bletchley Park had a big bonfire to burn all remaining evidence. Most of the bombe machines were destroyed, except for a few taken to a new secret location. Ultimately, even those machines were pulled apart and recycled.

Historians reckon that, through their work, the code breakers were vital for getting supplies across the Atlantic and shortened the war by two years. In doing so they will have saved lives, yet they were never soldiers. It is an example of brains over bullets.

---

[7] C. Hugh O'D. Alexander, 'Cryptographic History of Work on the German Naval Enigma'. The National Archives, Kew, Reference HW 25/1.

5　The Enigma of Emotion

TIFFANY WATT SMITH

What happens if you notice a person sitting next to you, and try to guess as to what emotions they are feeling? What would you look for? Their breathing? Their body language? The expression on their face? How easy is it to read the mysterious inner life of their feelings? When I started to think about the subject of this chapter, about enigmas on one hand and emotions on the other, two things happened. Firstly, I kept thinking about the experience of trying to detect from outward signs the private, inner life of a person's feelings. Secondly, a parade of people who are deliberately emotionally enigmatic came sauntering, nonchalantly, through my mind – James Bond, the Mona Lisa, The Fonz, poker-players, therapists, teenagers – people who seemed to make it intentionally hard to read how they are feeling. We live in a world that prizes the skill of reading how others are feeling and, in turn, increasingly expects us to make ourselves emotionally legible to one another. The latter demand falls more heavily on, and is riskier for, some people rather than others. This chapter focuses on what happens when we remind ourselves how difficult emotions are to recognise and name, while also considering the circumstances under which one might deliberately hide one's feelings as an intentional strategy of defiance or defence.

Confusing Feelings and Emotion Machines

Charles Darwin wrote *The Expression of the Emotions in Man and Animals* in 1872, at the height of his fame. It became a public sensation, a testament to the feverish interest middle-class reading Victorians had in their emotions. People sometimes say we are the first age to be fascinated by our feelings, but the Victorians undoubtedly got there first. Darwin's notebooks reveal that he became interested in emotions as early as the 1830s. Whilst on board the expedition ship *The Beagle*, he noted down the emotions he witnessed in the animals and people around him. When his first child was born in 1839, he immediately began a journal, noting the 'various expressions which he exhibited'. At that time, the prevailing explanation for human passions was theological, and contended that God implanted blushes and frowns, smiles and grimaces. Darwin's proposition was that the emotions themselves must have a 'gradual and natural origin', that they must have evolved to protect us from harm – such as disgust or fear – or to help us bond – such as love or compassion – and that, in turn, the modern emotional body must contain vestiges of our remote ancestors.[1]

In 1867, Darwin began work on emotion in earnest. His archives reveal that he corresponded with missionaries and explorers around the globe, asking them to describe the emotional characteristics of the indigenous people they encountered. He himself became a collector of emotions closer to home, going to photographers' studios and studying postcards of forlorn-looking children and histrionic, declaiming actors. He studied his own facial muscles in the mirror as he smiled and frowned. He observed his pets – his dog particularly, which was prone to disappointment when his walk was cut short ('the head drooped, the tail was by no means wagged'). And, as you might find yourself doing, Darwin stole glances at strangers, at dinner parties, in the street, noting the way emotions flitted across their faces. Darwin recalls an especially intriguing anecdote when, on a train journey, he found himself in a carriage sitting opposite a woman. He observed that 'her depressors *anguli oris* became very slightly, yet decidedly contracted', a muscular contraction which Darwin recognised usually preceded tears. The expression caused him to imagine 'some painful recollection, perhaps that of a long-lost child, was passing through her mind'.[2] So much is encapsulated in this 'perhaps'.

In fact, Darwin confessed on the opening pages of *The Expression of the Emotions* that '[t]he study of emotions is difficult' – as no doubt we have

---

[1] F. Darwin (ed.), *The Life and Letters of Charles Darwin*, 3 vols. (London: Murray, 1887), vol. 1, p. 113.
[2] C. Darwin, *The Expression of the Emotions in Man and Animals* (London: Murray, 1872), p. 193.

all experienced at times. The physiological differences are slight and fleeting and, moreover, he wrote, it is too easy to imagine you see the emotion of another when it might not be there, since your own mood infects any claim to objectivity.[3] In a way, *The Expression of the Emotions* is a book as much about the problem of observing emotions as it is about the emotions themselves. It is a book that plunges the reader vertiginously into the question philosophers call 'the problem of other minds'. It recognises that the relationship between the outward signs and symptoms of emotion and the mysterious, subjective inner feelings is not always a stable one, and that when we do try to 'read' another's feelings we can easily get it wrong.

It is nearly 150 years since Darwin published *The Expression of the Emotions*. Today, the idea that we should try to identify and name our own emotions and then learn to recognise – as much as we can – those of other people, is taken to be a public good. The school curriculum teaches children to label emotional states. People seeking psychological support are presented with a page of terms to try to improve their emotional fluency. Emotional intelligence is a skill taught in management seminars up and down the country. Much of this enthusiasm can be traced to Daniel Goleman's 1995 book *Emotional Intelligence*, which synthesised and popularised research conducted in the previous two decades. Goleman suggested emotional intelligence – presented as a learnable social skill – could be correlated with a number of positive outcomes, from making friends, to mental resilience, to success in business.[4]

We are a long way from becoming emotionally literate as a nation, of course. But a collective awareness of the role of emotion in shaping our decisions and behaviour is nonetheless growing. Emotions affect what we buy, the way we work, or do not work, how we vote. At the same time, we are repeatedly having to confront the reality that recognising other people's emotions – and often our own – is actually quite difficult, and so find ourselves open to the allure of so-called 'emotion-detection technologies'. Such technologies promise to put in the place of the fallible human

---

[3] Darwin, *Expression of the Emotions*, p. 13.
[4] D. Goleman, *Emotional Intelligence: Why It Can Matter More Than IQ* (New York, NY: Bantam, 1995).

mind – the mind of Darwin with his confession that observing emotions is 'difficult' – a silent, objective, reliable mechanical gaze. I am reminded of an episode of the American sitcom *The Big Bang Theory*. The sitcom features a group of socially inept scientist friends at Caltech. Sheldon, a brilliant but immensely annoying theoretical physicist, has difficulties recognising how others are feeling, which often lead him to offend and annoy. His friends discover that a team at MIT are developing a device that picks up subtle changes in heart rate and breathing to read human emotions. His friends suggest Sheldon ask for a prototype. 'A machine that reads emotions is intriguing,' he says, 'It could help make me a more considerate friend. It could also help me identify my enemies, discover their fears, and then I could use those fears to destroy them.'[5]

In 2019, *The Guardian* reported that so-called 'emotion machines', which allege to be able to identify anger, fear, disgust, sadness, and other emotions in faces in a crowd, has grown to a twenty-billion dollar industry.[6] These technologies are based on an assumption that the internal terrain of feeling can be detected by facial expressions, primarily muscular contraction around the eyes and mouth, founded on outdated research that suggested that there are a handful of basic emotional expressions – usually six – which are common to all people across the globe and can therefore be 'read' by a computer. The psychologist Paul Ekman has been very influential in this area, with his claim to be able to interpret 'micro-expressions', which, like a tell in poker, reveal even those emotions we do our best to hide. Many researchers since have taken issue with Ekman's conclusions, arguing that his photographic representations of the emotions do not, in fact, accurately represent all ethnicities and, in particular, lead to inaccurate conclusions with people of East Asian heritage.[7] In 2003, an attempt made by the US Transport Security Administration to

---

[5] M. Cendrowski, *The Big Bang Theory*, 'The Emotion Detection Automation', Season 10, episode 14, 9 March 2017.
[6] O. Schwartz, 'Don't look now: Why you should be worried about machines reading your emotions', *The Guardian*, 6 March 2019.
[7] Among the psychologists who have called Paul Ekman's FACS method into question are Rachael Jack and her colleagues. See R. E. Jack, C. Blais, C. Scheepers, P. G. Schyns, and R. Caldara, 'Cultural confusions show that facial expressions are not universal', *Current Biology*, 19 (2009), 1543–1548.

use emotion detection to detect potential terrorists was shown not only to have failed, but even to have simply re-rehearsed racial profiling.

Nonetheless, the emotion detection business is booming, spirited along by the belief that, with enough data, the algorithms will be able to crack the code of emotions. Rana el Kaliouby, the CEO of Affectiva and one of the leaders in this field, believes this technology will 'tap into our visceral, subconscious, moment by moment responses', using facial recognition software and a global repository of data gleaned from people watching TV, playing computer games and driving, to train algorithms to analyse the emotional content of faces in a crowd.[8]

Perhaps this makes you uncomfortable. It should do. And not just because of its Big Brother surveillance implications, or the risk of the technology falling into the hands of those who seek political power, or because a US government study showed last year that facial recognition algorithms have a significant racial basis – are far less accurate at identifying African-American and Asian faces than Caucasian faces – which makes their use by police and the courts dangerously irresponsible.[9] The idea of emotion detection is also problematic because it is based on an outdated and over-simplified notion of what an emotion really is. Professor Lisa Feldmann Barrett at Northeastern University is among a growing number of neuroscientists who have argued that the idea that emotions can be detected from simple facial expressions, or even from other physiological markers, is problematic. She argues that the brain is immensely flexible, wiring itself to whatever environment it develops in. There is, in her view, a dynamic relationship between language, culture, and brain chemistry, through which emotions emerge in highly complex ways, certainly involving automatic cognitive processes but also involving interactive, socially situated, and embodied experiences.[10] The picture she paints of emotions is suggestive of continually mutating forms, of plasticity rather than discrete and fixed

---

[8] Interviewed in Schwartz, 'Don't look now'.

[9] P. Grother, M. Ngan, and K. Hanaoka, *Face Recognition Vendor Test (FRVT). Part 3: Demographic Effects* (National Institute of Standards and Technology, US Department of Commerce, 2019), https://nvlpubs.nist.gov/nistpubs/ir/2019/NIST.IR.8280.pdf.

[10] L. F. Barrett, *How Emotions Are Made: The Secret Life of the Brain* (London: Macmillan, 2017).

physical signs, and describes a level of complexity that would be impossible for any machine – or human – to reduce to repeatable, predictable patterns. Through her writing, emotions emerge not so much as codes that can be cracked, but as reminiscent of Darwin's image of the 'entangled bank' with which he ends *The Origin of Species*, with its many species and kinds of plants and animals, from the flitting insects to the crawling worms; an elaborate picture of interdependence, forever mutating, forever in the process of change.[11]

Against this backdrop, then, it is valuable to think again about enigmas – which puzzle or confuse – alongside emotions. Later in this chapter, I will consider how being emotionally enigmatic might be a form of resistance or disruption in a world which insists on knowing about, training, and normalising your feelings. What are the possible consequences, and for whom, in moments when we refuse to comply with this demand to make our emotions legible? King Lear asks his daughters to tell him how much they love him, but Cordelia, precipitating her banishment and the tragedy, refuses: 'I can not heave my heart into my mouth.' Before approaching those deliberately enigmatic or withholding figures, we must first circle back round to the question *what is an emotion?* Because, if emotions are not simple reflex responses to external stimuli, encoded in our bodies at birth, then what are they?

What Is an Emotion?

People have been trying to understand their unruly passions in order to try to control them for a very long time, at least since the Stoic philosophers of the ancient world and their treatises on anger and love. As I mentioned earlier, Darwin's *The Expression of the Emotions* arrived at – and helped precipitate – a moment of intense interest in studying the emotions through a new scientific lens, one which privileged objective observable physiological signs, such as raised heartbeats and dilated pupils, over strange shadowy theological ideas. Even the word 'emotions' was still relatively new at this point, first introduced in the early decades

---

[11] C. Darwin, *On the Origin of Species by Means of Natural Selection* ..., 6th edn (London: Murray, 1872 [1859]), p. 429.

of the nineteenth century by the philosopher Thomas Brown, to replace the older theological language of the passions and affections of the soul, and to demarcate a new secular approach to the life of the mind.[12]

Following Darwin, a generation of experimental physiologists and neurologists began to become interested in studying the effects of emotions on physiology, with new technologies of photography and film, and machines for measuring reaction times. One of my favourite experiments was conducted by the Italian Angelo Mosso, who wanted to understand the effect of embarrassment on the rabbits he was studying, and noted that the blood vessels in their ears pulsated vigorously when they became aware of being watched, a response he called a rabbit's 'blush'. By the beginning of the twentieth century, this new approach to emotional life as a set of reflexes, responses to external threats and temptations, had been firmly established in the psychological sciences.

This physiological turn to emotions was not without its critics. Among them were those who feared this new knowledge would be used to control unruly populations. In Aldous Huxley's *Brave New World*, written in 1932, citizens must have monthly VPS (Violent Passion Surrogate) treatments. The Controller explains it thus: 'We flood the whole system with adrenin [an adrenaline-like substance]. It's the complete physiological equivalent of fear and rage. All the tonic effects of murdering Desdemona and being murdered by Othello, without any of the inconveniences.' 'But I like the inconveniences,' says the Savage. 'We don't', the Controller replies, 'we like to do things comfortably.'[13]

Concerns about this physiological turn also came from less expected quarters, from historians and philosophers, and later anthropologists, who were beginning to doubt that emotional experiences were shaped by physiology alone. In a field-defining essay published in 1941, the historian Lucien Febvre, co-founder of the Annales School of social history, argued that the application of contemporary psychological frameworks and diagnoses to the past was an insidious anachronism. He suggested that emotions – even so-called basic ones like rage or

[12] T. Dixon, 'Emotion: The history of a keyword in crisis', *Emotion Review*, 4 (2012), 338–344.
[13] A. Huxley, *Brave New World* (London: Vintage, 2007 [1932]), p. 211.

pleasure – might have looked and even felt very different in the past. We should not assume to know what 'rage' meant at Napoleon's time, or what its displays looked like. There is 'an abyss' he wrote 'between the morals and sentiments of the men of that age and ours'.[14]

Historians did not immediately respond to Febvre's challenge for them to chart the affective life of the past. Nevertheless, by the 1960s, a different discipline, anthropology, was engaging with a similar issue and drawing attention to the diversity of emotional languages found around the world. Anthropologists noticed a range of concepts which described unfamiliar emotional states – like *Awumbuk* used by the Baining of Papua New Guinea, which describes the inertia they feel when guests depart, or *Amae*, a Japanese word for the pleasures of sinking into the care of another, or *Liget*, a word used by the Ilongot who live in the jungles of Nueva Vizcaya in the Philippines to describe a distinctive combination of anger and grief.[15] Such studies suggested that these emotional languages might shape entirely different worlds of experience, not least because the values attached to emotions around the world were not uniform. For instance, within the cooperative culture of the Pacific islanders of Ifaluk, an emotion known as *Song* describes the indignation felt when you realise you have got less than your fair share, and this emotion is taken very seriously indeed. The anthropologists who first described it for Western readers remarked that in many Western cultures this emotion might be felt to be perhaps childish, or envious or petty, and might quickly be shut down. In Ifaluk culture, which prides itself on fairness and exists because of the intense cooperation of its members, *Song* is celebrated, and anyone who expresses it is immediately met with a flurry of people trying to put it right.

Encouraged by such insights, the historians got to work and began to unearth the forgotten emotional cultures of the past. As had the anthropologists, historians recognised that so much of what we feel is shaped by the cultures we live in. The fashions and moral judgements that surround us, the

---

[14] L. Febvre, 'La sensibilité et l'histoire', trans. K. Folca as 'Sensibility and history: How to reconstitute the emotional life of the past', in P. Burke (ed.), *A New Kind of History: From the Writings of Febvre* (London: Routledge & Kegan Paul, 1973), p. 25.

[15] For more on these and other 'untranslatable' emotions words, see T. W. Smith, *The Book of Human Emotions* (London: Profile, 2015).

religious rituals, and even what we think an emotion is can bring emotions to life and make them disappear again. Historians uncovered certain emotional categories that had disappeared – such as Acedia, a lethargy and sorrow which only descended around lunchtime and was first described by the desert-dwelling monks in the Sinai desert in the second and third centuries. They described other emotions which suddenly popped into being, a focal point for the anxieties of their time. Today we have invented the emotional category of FOMO, or fear of missing out. In the nineteenth century, Victorians coined another word: 'boredom', which first appeared in 1853 in Charles Dickens's *Bleak House*. It described the sensation of finding oneself in the midst of a world of diverting pleasures and yet being uninterested or at a loose end, a feeling that indexed both a moral failing and serious social threat. Historians also charted a forgotten landscape of unexpected physiological responses. In Chaucer's *Troilus and Criseyde*, written in the 1380s, the hero – a brave knight – faints in dismay. In the late twelfth century, the troubadour Arnaut Daniel described himself yawning for love. Historians additionally traced the way some emotions had changed their meanings, sometimes very dramatically, across the centuries. In the eighteenth and nineteenth centuries, Europe was gripped by an epidemic of Nostalgia, a fatal form of homesickness, an illness considered so grim and contagious that Swiss soldiers fighting abroad were forbidden to sing the folksongs of their homeland in case it brought on an attack. By the early twentieth century, not only did people no longer die of nostalgia, but the word had come to mean a pining for the past, rather than a longing for home.

These kinds of historical insights gave rise to a systematic study of emotions, with talk of 'emotional communities' governed by shared emotional values and rituals, and 'emotionologies', the rules which appeared in advice literature, medical, or legal texts, which instructed people about which feelings were desirable and which ought to be hidden. The assumption historians made, and with which I would agree to a large extent, was that rules such as emotionologies and emotional communities shaped, more or less, people's private experiences.[16]

---

[16] For an overview of these concepts, see J. Plamper, *The History of Emotions: An Introduction*, trans. K. Tribe (Oxford and New York, NY: Oxford University Press, 2015).

To give you an example from the history of portrayals of friendship in twentieth-century girls boarding school fiction. If you look at the works of early twentieth-century authors such as L. T. Meade, Angela Brazil, or Elinor Brent-Dyer, you will find that the friendships depicted are very romantic, a world of crushes and pashes. The girls almost swoon when they first set eyes on the girl who will become their friend, they hold hands, declare their love for one another, cuddle, and kiss, sometimes they get into bed with one another. This is very different from the model of friendship captured 40 or 50 years later in Enid Blyton's *Malory Towers* series, where a close friendship between two girls is still highly desirable, but the model is more pragmatic and self-sufficient. No one talks of loving one another, and the girls rarely even link arms except when celebrating a success on the lacrosse field, or plotting some terrible trick to play on the French mistress. Of course, this is fiction, and is usually interpreted as due to the personalities of their authors. Nevertheless, it is possible to see tangible change in emotional culture when we look at real girls' boarding schools. At Wycombe in the 1950s, for example, girls were forbidden to hold hands, walk arm-in-arm, or wash each other's hair; at around the same time in a teacher's training college, kissing was forbidden, except on birthdays.[17]

There may be several reasons for this change. Rosemary Auchmuty has argued it arose partly because of new anxieties circulating as a result of the new science of Sexology and its demarcation of a new category of so-called 'sexual inversion', i.e., lesbianism.[18] Perhaps it was also stoked by fears about mob mentalities and the desire to foster a spirit of self-sufficiency and independence that had come into stark focus during both of the world wars. School stories, viewed both as a source of entertainment and as moral education, reflected these new emotion display rules and the larger cultural shifts – in medicine, politics, and education – which gave rise to them.

One reason it is so important to pay attention to these shifts in emotional regimes and styles is that they remind us that emotions are

---

[17] G. Avery, *The Best Type of Girl* (London: Andre Deutsche, 1991), p. 306.
[18] R. Auchmuty, *A World of Girls: Appeal of the Girl's School Story* (London: Women's Press, 1992).

not stable, ahistorical, purely biological responses but are profoundly enmeshed in the cultures in which we live. History teaches us that emotions change, but does this help us solve the enigma of emotion? Could interpreting how a person is feeling simply be a question of putting together our physiological knowledge together with our understanding of display rules of different cultures? That certainly would, in my view, be an improvement on current ideas of emotional legibility, but of course nothing is really that simple. In her memoir *Fierce Attachments*, Vivian Gornick writes of the women in the tenement in the Bronx she grew up in: 'I absorbed them as I would chloroform on a cloth laid against my face.'[19] Sometimes, however, the cultures we live in do not work on us quite so powerfully, or completely. After all, you or I might know the rules, but that does not mean we follow them. We are human after all.

We all sometimes feel against the grain. Perhaps you get the giggles at a funeral or feel embarrassed by the lavish expensive dinner that everyone else seems to be enjoying. Perhaps, like Sarah Ahmed's 'feminist killjoy', you may greet your own wedding day not with uncomplicated joy – 'it's the happiest day of your life' say all the magazines – but with awkward feelings. Perhaps you are not frightened of dentists. Perhaps, when you get an award, you feel fraudulent rather than proud. Perhaps you feel panicked by the new baby rather than overwhelmed with love. Perhaps you feel weirdly euphoric at your approaching death, as did the guitarist Wilko Johnson when diagnosed with terminal cancer. Though we might inhabit the world, this does not mean its rules always entirely inhabit us. And, as I will explore shortly, sometimes we must disobey them to survive.

When we look at emotions – our own and other peoples – we see that there are physical reflexes and responses, on the one hand, and cultural rules, on the other, but neither of these entirely dictates how and why we feel the way we do. Shaping emotions is also a vast and ever-shifting network of individual experiences and histories, of personal tastes and desires, of the emotional communities we belong to, not only the publicly accessible ones, but also those which are private and belong to families, to

---

[19] V. Gornick, *Fierce Attachments: A Memoir* (New York, NY: Farrar, Straus and Giroux, 2005), p. 4.

couples, to friendship groups. This is such a vast complex that it would be meaningless to attempt to reduce it to shared or universal principles or categories which allow us to detect this rich inner life of feeling from outward signs.

Trying to categorise and 'read' emotions is a bit like John Ruskin's attempts to master the language of the sky. He advised students to set aside 15 minutes a day to study the sunrise, keeping a pencil and paper to hand to sketch the melting, shifting forms of the clouds. Yet, just as Darwin had described the difficulties of studying emotions, so Ruskin lamented the difficulty of capturing the clouds. His drawings were 'little more than memoranda of the skies as they passed', their categories more a matter of convenience than true description.[20] What is more, within this strangeness, this emotional illegibility, there is the potential for hiding and evading, and so defying the emotional expectations of the worlds we live in.

The Art of Inscrutability

Recall, if you can, the Mona Lisa and her famously inscrutable smile. How many words have been dedicated to this smile? Is it the consequence of sexual allure, of bad dentistry, of a ghastly secret? We seem invested in her as an enigma, even enjoy being diminished by her powerful gaze, because by not giving away her feelings, by playing her cards close to her chest, she retains her status and her enigmatic allure. Leonardo da Vinci knew this. By depicting some feeling, but not making it easily readable, he makes us lean in, entranced.

In a world of emotional declarations, transparency, and legibility, it is worth pausing for a moment to think about the emotionally enigmatic; not only because, cool and aloof, they are strangely alluring individuals such as James Bond or Tilda Swinton. Withholding a display of emotion can be considered its own 'emotional labour', a distinctive aspect of the work of police officers and judges, among others, who need to conjure the

---

[20] J. Ruskin, 'Notes on educational series' (1870), in E. T. Cook and A. Wedderburn (eds.), *The Works of Ruskin*, Library Edition, 39 vols. (London: George Allen, 1903–1912), vol. 12, p. 106.

appearance of objectivity and resistance to being easily moved. While doctors may need to embody a stoic calm in the midst of their patients' tempests, therapists and others in psychological caring professions deliberately make themselves into the famous psychoanalytic 'blank screen' to make space for the emotions of the person they are listening to, but also to invite the patient's transferences and projections, and hold them up for inspection.

As many artists know, there is a freedom that comes with the emotionally enigmatic, the freedom to impose your own narrative – as did Darwin with the woman who sat opposite him on the train – or intensify your desires. I am thinking here of the ending of the 1933 film *Queen Christiana*, which stars Greta Garbo. The camera pans in on a close up of Garbo's face as she stands at the figurehead of a ship bound for Spain. As they were filming, Garbo asked, 'What do I express in this last shot?' and the director Rouben Mamoulian's answer was 'Nothing. Absolutely nothing. You must make your mind and your heart a complete blank.' The resulting image is a face that, as Judith Brown writes, 'is not reducible to trivial emotion but instead remains mysterious, unreadable, yet entrancing to its viewer'.[21] Mamoulian wanted his audience to write the ending of the film themselves, and the resulting image produces, as Roland Barthes put it, the kind of exhilaration and ecstasy which comes when witnessing an image of the divine.[22]

Inscrutability can also produce less poetic responses. It can back people into a corner, creating feelings of awkwardness, confusion, bafflement, even rage. Here, I turn to a very different film, the Coen Brothers' *Miller's Crossing* from 1990, a gangster film set during Prohibition. To prove his loyalty, Tom Reagan is commanded to kill a friend, Bernie, in the woods at Miller's Crossing. The resulting scene, which lasts three-and-a-half agonising minutes and is justly famous, sees Tom, played by Gabriel Byrne, silently walking Bernie at gunpoint deeper into the woods. When the camera swings close-up into Tom's face, we see its expression remains completely motionless, for the entire three-and-a-half minutes, his eyes

---

[21] J. Brown, quoted in J. Brown, 'Garbo's glamour', in A. Jaffe and J. Goldman (eds.), *Modernist Star Maps: Celebrity, Modernity, Culture* (Farnham: Ashgate, 2010), pp. 107–122 (p. 113).
[22] Roland Barthes, *Mythologies* (London: Vintage, 1993), p. 56.

almost hidden by the shadow of his wide-brimmed hat, his lips like a pencil with both ends sharpened. Even his body language – his walk slow and even, his gun held steady in his hand – is impenetrable. The tension becomes unbearable for the viewer, and for Bernie, whose desperation grows the longer Tom refuses to answer his entreaties, the silence ultimately makes him vomit up his pleading in stuttering and wails.

I want to turn now to think about two extremes of emotional ambivalence and the effects produced by them. At one end of the spectrum is the presentation of excessive or confusing emotion that is made hard to read. On the other is the aura of cool nonchalance and emotional aloofness where the 'real' feeling is either masked or even inaccessible to the person feeling it themselves. Examining these ambivalent stances draws out the ways emotional illegibility might be used tactically, not so much as a way of grabbing power as a mode of defending yourself in a hostile world. First, then, I shall consider confusing or excessive emotion and, to do so, shift the scene to London, in December 1932, at a house in Holland Park Avenue, where a drag ball is in full swing, about to be raided by police.

Albert A was powdered, rouged, and bewigged, and wearing a feather headdress. Joseph C had on a low-cut evening dress and carried a handbag. As the police swarmed, lipsticks were hastily dropped, and the men rubbed the powder from their cheeks. The party's host, known to his friends as Lady Austen, was in the midst of being questioned by Inspector Frances, who was leading the raid, when he pointed to another party-goer, a man wearing a dress, and asked if that man had, in fact, been an undercover police officer. The inspector admitted he had. 'Fancy that', Lady Austen said, perfectly in control, his voice poised dangerously between irony and sincerity, 'he is too nice. I could love him and rub his Jimmy for hours.'[23] This account comes from metropolitan police archives and appears in historian Matt Houlbrook's brilliantly researched book *Queer London.* The book charts a flourishing homosexual subculture in London in the period between 1918 and 1957, and police attempts to control it, even to the extent of going undercover in drag to balls. The night of the raid at Holland Park Avenue, 60 men were arrested and, early the following year,

---

[23] M. Houlbrook, *Queer London: Perils and Pleasures in the Sexual Metropolis, 1918–1957* (Chicago, IL: University of Chicago Press, 2006), pp. 244–245.

the case was sensationalised and widely reported: 'MEN DRESSED AS WOMEN', screamed the *Morning Advertiser*.

Lady Austen's retort – 'he is too nice. I could love him and rub his Jimmy for hours' – is quite moving. Here was a man who was under arrest and being questioned, who would be charged, and whose name would appear in the papers with the purpose of exposing and humiliating him. Yet, despite all that looming, he remained defiant, kept his poise, and delivered his ironic, slightly baffling, mocking reply: 'fancy that'. Lady Austen not only broke the expected emotional rules of such an encounter – arrests are more usually connected to fear, shame, or anger – but one imagines his retort might have wrong-footed the Police Inspector, confusing and humiliating him.

Susan Sontag, in her famous essay on camp, approached camp as sensibility and as an aesthetic, defined by its travesties, theatricality, and stylistic playfulness, and most of all its evasion of strong feeling.[24] In homosexual subcultures in 1930s London, camp was also a weapon. Homosexual life in the 1930s may have been more vibrant and visible than usually imagined, but it was still replete with dangers, a place where a jokey catcall might turn violent. The linguistic anthropologist Stephen O. Murray has argued that among certain subcultures – he studied African-American and Queer subcultures in the 1960s and 1970s – insults were part of self-protection. By this time, ritualised insult exchanges were known in Latinx and African-American Queer subcultures as 'throwing shade' and 'reading', or its danced form 'vogueing'.[25] These, writes Murray, provide 'entertainment during "dead time" and also practice for dealing with hostility'.[26] Alex Purdie, born in 1913, was a Deptford market trader with a flamboyant persona who wore makeup and had a string of boyfriends. He recalled how 'in a pub you get the [kiss sound] and all this lark ... you've got to be aggressive ... give them a mouthful'. As Houlbrook writes, 'to outsiders, this kind of response was unnerving,

<hr/>

[24] S. Sontag, 'Notes on camp' (1964), repr. in *Against Interpretation and Other Essays* (London: Penguin, 2009), pp. 275–292.
[25] See J. Livingston, *Paris Is Burning*, film (Academy Entertainment, 1990).
[26] S. O. Murray, 'The art of gay insulting', *Anthropological Linguistics*, 2 (1979), 211–223 (p. 211).

simply because it was so unintelligible'.[27] We do not know if the Police Inspector who arrested Lady Austen was baffled or even mildly embarrassed by his camp report with its confusing, ironic expression of desire and disappointment, but I cannot help feeling it might have taken a little of the wind out of his sails and, in so doing amidst the chaos and fear of 60 men being arrested, disrupted, for a moment, the distribution of power.

So much for the person who displays their feelings in an ambiguous or confusing manner, but what of the person who goes to the opposite extreme and withdraws their emotional display altogether, leaving a confusing flatness? What of their retreat into flatness? We move now to the law offices of New York's Wall Street, as depicted in Melville's story 'Bartleby the Scrivener'.[28] Bartleby is a silent, pale, and mechanical clerk to a successful lawyer who narrates the story. The story is well known (there are even t-shirts printed with Bartleby's famous catchphrase). When the narrator asks Bartleby to perform a simple task, he finds himself astonished by the man's reply: 'I would prefer not to.' The narrator is baffled: 'I sat awhile in perfect silence, rallying my stunned faculties.' He tries again, but the answer is the same: 'I would prefer not to.' 'Had there been the least uneasiness, anger, impatience or impertinence in his manner ... I should have violently dismissed him from the premises', he tells us. Yet, there is no obvious emotion in Bartleby, only we learn a sort of impassivity that feels like a 'strange wilfulness', a 'cadaverous gentlemanly nonchalance' that has the peculiar effect of rendering the narrator powerless, and emasculated: Bartleby 'not only disarmed me, but unmanned me', he confesses.

This kind of enigmatic emotional flatness has long been characterised as pathological. For Melville, Bartleby's withdrawal – what Melville calls his 'passive resistance' – is framed first as loneliness and, finally, as madness. 'I think he is a little deranged', the narrator says. To nineteenth-century readers Bartleby's response might have been recognisable as a case of aboulia, a type of mental illness that was characterised

---

[27] Houlbrook, *Queer London*, p. 150.
[28] H. Melville, 'Bartleby the Scrivener', repr. in *Billy Budd, Bartleby, and Other Stories* (New York, NY: Penguin, 2016), pp. 17–54.

by a complete retreat from, and loss of will and interest in, the world. Eventually, Bartleby refuses to leave, and the police are called. Bartleby is arrested for vagrancy and taken to the Tombs, where he refuses to eat – 'I would prefer not to' – and eventually is found dead huddled at the foot of a wall.

Aboulia was not the only way in which nineteenth-century observers made apparent emotional withdrawal 'other'. As Xine Yao, who works at the intersection of critical race studies and the history of emotions, has shown, accusations of being unfeeling also adhered to many non-Europeans at this time. She examines the cliché of 'the inscrutable Oriental' which circulated widely in nineteenth-century literature and ethnographies, which figured in stories about cheating gamblers and enigmatic prostitutes, or in ethnographies as representatives of a 'naturally taciturn' race.[29] Such clichés and pseudo-scientific claims shored up suspicions of insincerity, secretiveness, or hard-heartedness by turns. Yao argues for a different reading of these disaffected emotionally illegible figures, recognising the necessity that, in some cases, appearing 'unfeeling' is a deliberate strategy of surviving in a world all too eager to require emotional responses and then condemn them. Yao reminds us of Audre Lorde's open letter to Mary Daly written in 1979, in which Lorde describes the necessity of withdrawing from the emotional demands of others, in particular, withdrawing from the guilt and projections of white women: 'I had decided never again to speak to white women about racism. I felt it was wasted energy because of destructive guilt and defensiveness, and because whatever I had to say might better be said by white women to one another at far less emotional cost to the speaker.'[30]

I wonder whether the survival tactic of active emotional refusal, which Lorde so brilliantly articulates in relation to the power dynamics of racism and feminism, may be read to work more broadly in any situation where the cost of emotional transparency might be particularly high and

[29] X. Yao, *Disaffected: The Cultural Politics of Unfeeling in Nineteenth Century America* (Durham, NC: Duke University Press, 2021).
[30] A. Lorde, repr. in *History Is a Weapon*, www.historyisaweapon.com/defcon1/lordeopenlettertomarydaly.html.

land on the speaker unfairly. For this purpose, I will turn to a story first published in *The New Yorker* in 2017, which quickly went viral. The story was written by a fairly unknown – at that time – writer, Kristen Roupenian, and is called 'Cat Person'.[31] The story is about a college student, Margot, who has a job in a cinema where she meets Robert, a man in his early thirties. Robert, Margot tells us, 'was cute. Not so cute that she would have, say, gone up to him at a party, but cute enough that she could have drummed up an imaginary crush on him if he'd sat across from her during a dull class.' Right from the start, Margot emerges as a young woman whose low-key emotional world is often marked by moods of ambivalence and boredom. She is enigmatic even to herself, more at home with the concept of an imaginary crush than with a real one with its biting passion.

Their initial conversations unfold by text-exchange of deadpan jokes – 'they never talked about anything personal' – and their flirtation evolves into a date, to which Margot wears baggy leggings and a t-shirt, and during which Robert is cold and quiet. 'Glad to see you dressed up for me,' Robert says. She assumes it is a joke, 'but maybe she actually had offended him by not seeming to take the date seriously enough, or something'. Written in the very close third person, in a flat style that lets the narrative scaffolding fall away and reads more like a personal essay or diary entry than a story, we have privileged access to Margot's unsure, oscillating mind, but also to her sense of Robert as a man who is physically imposing, easily offended, and potentially retaliatory. They go back to Robert's house at Margot's instigation and are on the verge of having sex when Margot realises that she does not want to continue, that she is disgusted by Robert and that the sex will involve no pleasure for her. Yet 'the thought of what it would take to stop what she had set in motion was overwhelming; it would require an amount of tact and gentleness that she felt was impossible to summon'. After the sex is over, she thinks, with a bright irony, 'this is the worst life decision I have ever made! And she marvelled at herself, at the mystery of this person who'd just done this bizarre, inexplicable thing!'

---

[31] K. Roupenian, 'Cat Person', *The New Yorker*, 4 December 2017.

After Margot leaves, she does not willingly see Robert again. She knows that to ghost him would be 'inappropriate, childish and cruel', but she cannot bring herself to reply to his texts. Instead, she finds herself trapped by the exhausting job of trying to compose the perfect way to dump him without hurting his feelings. Eventually her friend snatches her phone off her: 'Hi im not interested in you stop textng me' [sic]. When he shows up at her student bar, she runs away without talking to him. She does not reply when he texts her later. On the one hand, Margot's withdrawal is immature and hurtful. And part of the attention the story got focused on how unpleasant both the central characters are. Surely, the emotionally intelligent response would be for Margot to face Robert, to have a clear conversation with him, be clear about her feelings and so on. Yet, Margot seems to intuit some risk in telling Robert how she feels, even the polite version. Just as she is overwhelmed by the thought of summoning up enough tact to stop the sex, finding a tactful way to dump Robert is paralysing. What Roupenian describes is that poisonous cocktail of having to manage another's feelings, especially when we know that effortful emotional work comes at a risk. We are repeatedly told that Robert is physically imposing, and easily hurt, and that these things together scare Margot. When we talk about emotional intelligence and the assumption that emotional legibility is of uncomplicated value for all, we forget to recognise the way that the cost of revealing emotions is greater for some than for others. It turns out that Margot's intuition was right. The story ends with a string of texts sent by Robert, ending finally with a single word: 'whore'.

Thinking about emotions and enigmas from several different perspectives adds to our understanding of the former in two significant ways. Firstly, although we know a great deal about emotions – about their physiology and about their cultures – they remain extremely complex and particular phenomena, not just varying greatly between historical periods or geographical locations, but also differing from person to person. We must treat with a healthy dose of scepticism any claim that an app or machine might be able to accurately detect or read emotions. Secondly, although we live in a culture which increasingly assumes a norm in which emotional transparency and legibility are both desirable and healthy, the demand that we make our emotions legible to others is not experienced similarly by all people in all circumstances. There are moments, encounters, and circumstances – many more than I have touched on here – when being emotionally illegible, whether confusing or withdrawn, may be a radical form of self-protection and a peculiar, but effective, mode of defiance.

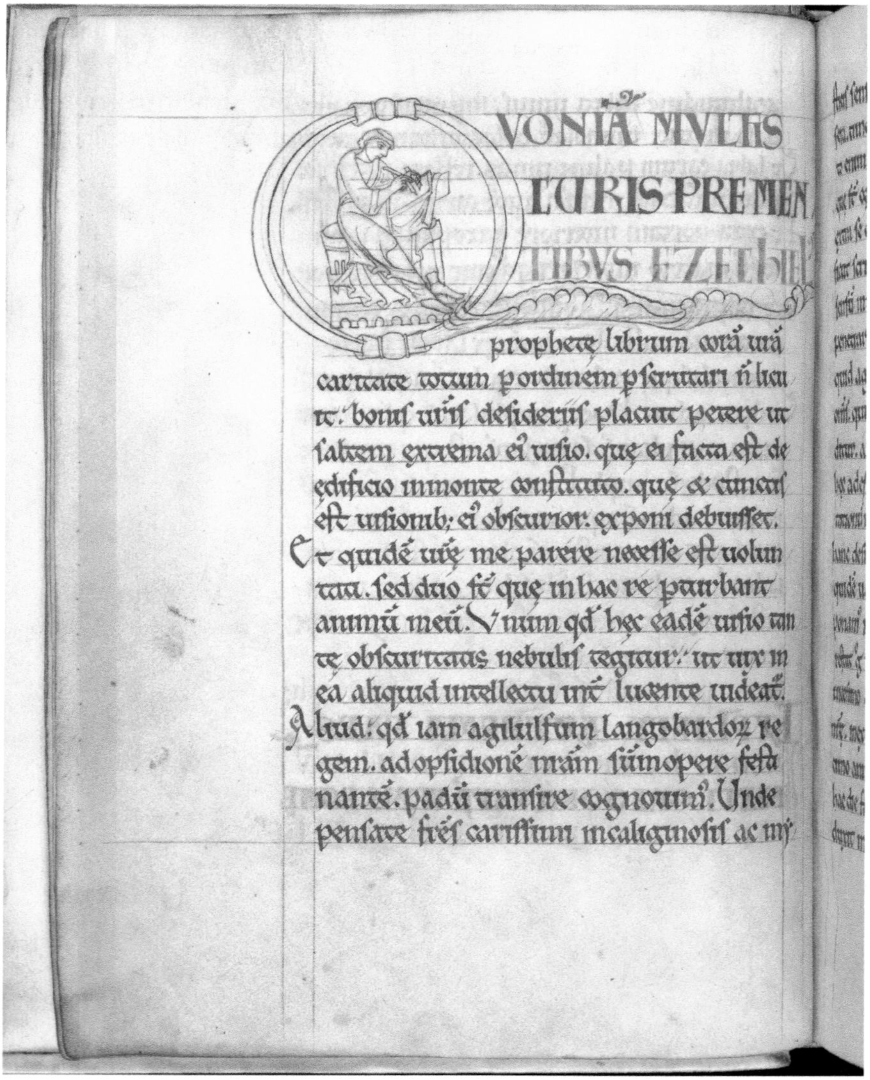

FIGURE 6.1 Manuscript from Rochester Cathedral Priory, 1100–1125 (London, British Library, Royal 4 B.i, fol. 4v), www.bl.uk/catalogues/ illuminatedmanuscripts/record.asp?MSID=5917.

6 The Enigma of Medieval Letters

ERIK KWAKKEL

The defining feature of the medieval manuscript, the premodern book at the heart of this chapter, is that it was written by hand rather than being printed (Figure 6.1). Unlike printed books, manuscripts were produced one at a time, and each was intentionally created for a specific reader and purpose. Scribes made sure that the book adhered to the preferences of the reader who had instigated its production, whether the object was made for personal use by the scribe, for a paying customer, or for the members of the religious community to which the scribe belonged. As a result of this close affiliation between producer and reader – a dynamic which was lost in the age of print – surviving medieval books carry information beyond that which is contained in the text on their pages. This makes the medieval book both an exciting and an enigmatic object of study: through the specific design preferences of its first reader, the manuscript offers significant clues as to the cultural or professional background of the reader, or to the purpose for which the book was made.[1]

The feature at the heart of this chapter, the configuration of the manuscript's handwritten letters, is arguably the most impactful design element that preserves historical information about the book itself. In fact, as will become clear, medieval letters produce two kinds of meaning. The first is that of the words they form: this kind of meaning is arguably the primary reason for consulting a manuscript or any book, then and now. The other kind is hidden within the material shapes of

---

[1] E. Kwakkel, 'Decoding the material book: Cultural residue in medieval manuscripts', in M. Van Dussen and M. Johnson (eds.), *The Medieval Manuscript Book: Cultural Approaches* (Cambridge: Cambridge University Press, 2015), pp. 60–76.

the letters that form the words. As they produced a manuscript, scribes incorporated information about the reader and the book's purpose into the manuscript's design, and through their handwriting they also embedded coded information about themselves. How scribes configured individual letters depended on a range of contextual variables, including the location and moment of production, the institution or monastic order in which the manuscript was produced, and the amount of care with which the text was copied. Each of these factors impacted the shape of the letters, but the moment and location of production was the most influential and will be the primary focus here. What information do letters carry beyond the words they form, and how can we tap into this source of information?

The Art of Studying Letters

There is nothing quite like handling a medieval manuscript for the first time. As you approach the library, anticipation rises. What will the book look like? How tall will it be? What page design did the scribe choose to use? Are any names of previous owners jotted on the flyleaves? In the digital age, some of these questions may already have been answered before arriving in the library, simply because so many medieval manuscripts have now been digitised.[2] Even with prior knowledge, however, meeting a manuscript in the flesh for the first time is special. The sweet sensation of discovery is never far away. This feeling is amplified by the fact that studying a manuscript is all about touch, about holding an object that was made perhaps a millennium ago and which has numerous written traces of earlier readers, all of them long gone. Encountering the inky fingerprint of a scribe is a powerful experience, as is seeing your thumb rest on the dirty corner of the page, precisely where dozens of medieval readers placed theirs. The Dutch historian Johan Huizinga posited that touching historical objects, especially medieval ones, can cause the sudden realisation of being one with the past, which he calls the 'historical sensation'. As the temporal gap between the researcher and

---

[2] See, for example, the exemplary library of digitised manuscripts in the Virtual Library of Switzerland (www.e-codices.unifr.ch/en).

the object closes, Huizinga states that 'a brief moment of drunkenness' may be experienced.[3] Being a scholar of medieval manuscripts comes with unexpected pleasures.

Opening a medieval book activates the senses. The parchment feels smooth to the touch, the binding utters a muffled squeak as the front board is gently lifted, and the resulting movement of air carries a musky smell from deep within the book. Then, the eyes get busy. When the first text page appears, the eyes of the palaeographer, the specialist of medieval handwriting, start to hover over letters. It is in this moment that something remarkable happens. The letter shapes begin to prompt unsolicited judgements regarding the moment and location of the manuscript's production, and whether the scribe was accomplished or not, or in some cases even regarding the identity of the person who produced the manuscript. Remarkably, such realisations are produced in a heartbeat. If one could look at the process in slow motion, one would observe the palaeographer's eyes pause, take in specific letter forms, and register their shape, from the height of ascenders and the angle of pen strokes, to the shape of the loops, bellies, and feet of letters. Equally remarkably, palaeographers retain these letter shapes in their heads. The details of each observed specimen of handwriting are stored in the expert's mental repository, where the letter forms and their component traits sit among thousands of impressions derived from other manuscripts. In this vast collective of mental images, the mind's eye starts to observe similarities and patterns – and to make connections. This process allows the palaeographer to make intuitive calls regarding the book's time and location of production, or the identity of its maker. It is an unbeatable sensation: to *know* that you know something about the origins of a manuscript, even if the book is as silent as the animals from which it was once made.

It is notoriously difficult to substantiate palaeographical verdicts regarding a manuscript's age or location of production, and it is with this issue of evidence – or rather, the lack thereof – that the present chapter is concerned. Medieval script experts tend to favour 'soft'

---

[3] J. Huizinga, *Verzamelde werken: Deel 2* (Haarlem: Tjeenk Willink, 1950), p. 564 ('sensation') and p. 566 ('drunkenness'). See also F. R. Ankersmit, *De historische ervaring* (Groningen: Historische Uitgeverij, 1993), p. 11.

evidence, expressing, often with great precision and conviction, how they perceive the handwriting of a given scribe. In confirming the date of production, the impression produced by the letter forms is more important to many palaeographers than the precise shape of the letters themselves. In an attempt to quantify these impressions, experts may note that the 'boldness' or 'vigour' of a script matches the proposed region or moment of creation, not unlike sommeliers arguing for a wine's vintage. In palaeography the notion of 'aspect' has come to cover such associative, intuitive verdicts. The influential British palaeographer Malcolm Parkes defined aspect as 'the general impression on the page made by a specimen of handwriting at first sight'.[4] The term is particularly useful for engaging with the enigma of meaning hidden in a medieval letter's shape. The opaqueness of aspect reflects the intuitive and subjective process behind palaeographical determinations. Moreover, the term recognises the importance of 'general impression' in the process of making such assessments. It is not usually a specific palaeographical feature – an atypical shape or flick of the pen – that connects the palaeographer to the hidden layer of information inside letter forms. Rather, it is the collective impression generated by all of the letters combined that creates the sensation of a certain date and location of production, or even the identity of the scribe. The process has been likened to seeing a letter from a loved one from a distance: '[J]ust as any of us could recognize our mother's handwriting from crossing a room in which her letter lay on the desk, a scholar who has become familiar with a scribal hand can establish a likelihood of its identity in the first instance by a quick glance.'[5] What sparks this awareness, however, is not usually clear; in practice it may be difficult to explain what made you realise who sent the letter.

In spite of the challenges inherent in their evidence, palaeographers are able to date medieval manuscripts with an increasing level of precision. And it is important that they do so, because scholars in history, art-

[4] M. B. Parkes, *Their Hands before Our Eyes: A Closer Look at Scribes, The Lyell Lectures* (Aldershot and Burlington, VT: Ashgate, 2008), p. 149 ('aspect').

[5] L. R. Mooney, 'Professional scribes? Identifying English scribes who had a hand in more than one manuscript', in D. Pearsall (ed.), *New Directions in Later Medieval Manuscript Studies: Essays from the 1998 Harvard Conference* (York: York Medieval Press, 2000), pp. 131–141 (p. 136).

history, and the textual disciplines lean heavily on the outcome of palaeographical research, even if solid evidence is often lacking. In the absence of the title page, an invention of the printing age, handwriting is the main source for determining where and when a medieval historical or literary text was copied. The only other source for this information is the scribal colophon, the medieval precursor of the title page. Colophons can be quite informative, not just about the date of production, but also about the circumstances under which the manuscript was written. On the last page of a manuscript in Leiden University Library, the scribe states that 'This book was written in the year 1484. It was completed on St Maurice's Day in the city of Susteren, where we were in hiding after our convent had been burnt down.'[6] As revealing as colophons can be, relatively few provide such clear details of when and where the scribe wrote the book. Moreover, colophons are encountered in less than 5 per cent of surviving manuscripts.[7]

The low frequency rules them out as a key instrument for dating and localising medieval books, while underscoring the importance of letter forms in making these determinations. However, before turning to medieval letters and what they communicate about a manuscript's production, our focus will briefly shift to a related discipline that deals with such concealed contextual evidence: codicology (from 'codex', meaning 'medieval book'), or the study of how the manuscript was put together.[8] There is much to learn from codicological methods as we calibrate our instrument for tapping into the information hidden inside a scribe's handwriting.

The Science of Studying Letters

Codicological studies show that it takes a systematic study of hundreds of specimens to draw out the historical significance of a manuscript design

---

[6] Leiden, Universiteitsbibliotheek, BPL 2541, fol. 347v. See E. Kwakkel, *Books Before Print* (Leeds: ARC Humanities Press, 2018), pp. 68–69.

[7] E. Overgaauw, 'Where are the colophons? On the frequency of datings in late-medieval manuscripts', in R. Schlusemann, J. M. M. Hermans, and M. Hoogvliet (eds.), *Sources for the History of Medieval Books and Libraries* (Groningen: Egbert Forsten, 1999), pp. 81–93.

[8] J. P. Gumbert, 'Fifty years of codicology', *Archiv für Diplomatik, Schriftgeschichte, Siegel- und Wappenkunde*, 50 (2004), 506–526.

feature, but also that doing so provides the kind of empirical evidence that is so rarely encountered in palaeographical studies.[9] The field of quantitative codicology has tackled such issues as the size of the quires, showing, for example, how certain numbers of double-leaves per quire were favoured in certain times and geographical locations.[10] Such insights are important because they help determine where and when medieval manuscripts of unknown origins were produced. Codicological assessments can be surprisingly precise. Consider the phenomenon of how manuscript producers engaged with the top line of a page. Before they commenced writing, scribes ruled the page with the help of a ruling device, namely a sharp object, a pencil, or a pen. In the 1950s, the British palaeographer Neil Ker noticed how some scribes wrote on top of the first ruled line on the page ('writing above topline', as visible in Figure 6.3 later in this chapter), while others started on the second line ('writing below topline'). On the basis of the manuscripts he had examined, Ker inferred that it concerned a transition that took place in the early thirteenth century.[11] A more systematic study from 1988, based on dated and datable manuscripts, confirmed his inference and determined that the transition took place quite suddenly *c.* 1240.[12] In other words, the scribe's treatment of the topline helps determine whether a thirteenth-century specimen of writing is more likely to belong in the first or in the second half of the century.

Quantitative codicological studies are generally based on manuscripts with dated colophons, which provide evidence that is firmly anchored in time and geographical space. To facilitate this kind of research, production of a series of catalogues with dated and datable manuscripts, entitled

---

[9] Gumbert, 'Fifty years of codicology', 505–507; E. Ornato with contributions from his friends and colleagues, *La face cachée du livre médiéval: L'histoire du livre*, I libri di Viella 10 (Rome: Viella, 1997), with an analysis of the quantitative method at pp. 1–83.

[10] P. Busonero, 'La fascicolazione del manoscritto nel basso medioevo', in P. Busonero, M. A. Casagrande Mazzoli, L. Devoti, and E. Ornato (eds.), *La fabbrica del codice: Materiali per la storia del libro nel tardo medioevo* (Rome: Viella, 1999), pp. 31–135.

[11] N. Ker, 'From "above topline" to "below topline": A change in scribal practice', *Celtica*, 5 (1960), 13–16.

[12] M. Palma, 'Modifiche di alcuni aspetti materiali della produzione libraria latina nei secoli XII e XIII', *Scrittura e Civiltà*, 12 (1988), 119–133. An English translation, 'Changes of some material aspects of twelfth and thirteenth-century Latin book production', has been made available by G. Bevilacqua, www.academia.edu/5529467/CHANGES_OF_SOME_MATERIAL_ASPECTS_IN_THE_TWELFTH_AND_THIRTEENTH-CENTURY_LATIN_BOOK_PRODUCTION_Marco_Palma_.

the *Catalogues des manuscrits datés*, was started in the 1950s. Now comprising over 40 volumes, it includes thousands of manuscripts that feature a scribal colophon, each represented by a succinct description and a photograph.[13] Manuscripts with dated colophons are indispensable for tracing changes in the mannerisms of medieval scribes, because they help connect a scribal practice to a particular period and geographical location, especially when they are studied in a quantitative manner. However, while the *Catalogues des manuscrits datés* have prompted numerous inquiries into the design aspects of the medieval book, few quantitative *palaeographical* studies have been undertaken – and for good reason.[14] The effort of using script for statistical research hinges on observing and quantifying subtle differences in the configuration of letters, which then must be captured in objective, quantifiable terms. It is hard to express how two presentations of the letter **s** are different – much harder than, for example, reporting whether a scribe started writing on top of the first ruled line or below it. Even more difficult than the quantification of individual letter forms is making verifiable claims about a *style* of writing (comprised by all letters of the alphabet, the abbreviations, and the punctuation), which is why the term 'aspect' is so ubiquitous: while the expert senses in a flash that a script is 'bold' or 'pointy', defining in objective terms precisely what makes it bold or pointy is far harder.

Strokes

A quantitative approach to medieval script starts by defining in what ways two representations of the same letter are or can be different. To do so, criteria are needed that capture, in an objective manner, the minute variations in the configuration of letters. These variations, many of which become apparent only when enlarging digital images, are responsible for the overall 'look' of a specimen of handwriting, and it is this look that

---

[13] 'État des publications', CMD: Catalogues de Manuscrits Datés, www.palaeographia.org/cipl/cmd.htm.

[14] A. Derolez, 'Possibilités et limites d'une paléographie quantitative', in P. Defosse (ed.), *Hommages à Carl Deroux. V: Christianisme et Moyen Âge, néo-latin et survivance de la latinité*, Collection Latomus 279 (Brussels: Peeters Pub & Booksellers, 2003), pp. 98–102; some quantitative studies are mentioned at p. 102, n. 14.

subsequently prompts the sensation that something originates in the twelfth century and was made in France, rather than in Germany a century later; or that a specimen has been written in a bold or pointy style. The key to tracking and defining these subtle script variations is found in the manner in which a scribe inscribed a letter on the surface of the paper or parchment page, a process called *ductus* (duct), or '[t]he act of tracing strokes on the writing surface'.[15] As it happens, stroke variation makes it possible to measure and to count features of letters in an objective manner, and with the same ease as mapping codicological mannerisms such as writing below or above the topline. The process starts by precisely registering how a letter is formed.

When the palaeographer's eyes travel across the page, the handwriting of the scribe generates questions that may appear peculiar to the non-expert. What do the arches of the letter **m** look like; are they round or more angular? Is the foot of the letter **f** placed on the line, or is it extended below it? Does the ascender of the letter **d** resemble a broomstick or a baseball bat? Such questions originate from a consistent and important feature of medieval handwriting – that it evolved over time. In the twelfth century, the letters **d**, **f**, and **m** were shaped very differently from how they appeared in the ninth century. Even within each century, letter shapes were not consistently the same. To make sense of this variance and development in medieval handwriting (and, crucially, to teach it to students), palaeographers distinguish different script 'families', as they can be called: writing styles with shared and clearly identifiable features, which are commonly related to specific time periods and sometimes to specific geographical areas. In palaeographical handbooks each family is given a name and described as a separate entity, producing a tidy – though highly oversimplified and not necessarily historical – reconstruction of the medieval writing system.[16]

---

[15] Parkes, *Their Hands before Our Eyes*, p. 151 ('ductus').
[16] The main book scripts of the medieval period are described in A. Derolez, *The Palaeography of Gothic Manuscript Books: From the Twelfth to the Early Sixteenth Century*, Cambridge Studies in Palaeography and Codicology 9 (Cambridge: Cambridge University Press, 2003), with special attention paid to chronological development and regional variation in the shape of letters.

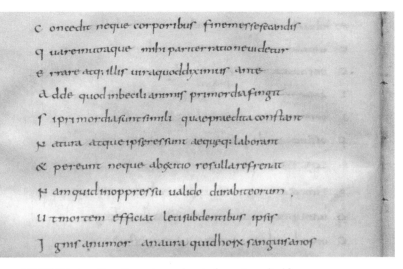

FIGURE 6.2 Caroline minuscule, ninth century (Leiden, Universiteitsbibliotheek, VLF 30, fol. 22v, detail).

Thus, we are taught that Caroline minuscule (Figure 6.2) is the main script of the Early Middle Ages. It came into being in the late eighth century and remained in use until *c.* 1100, when it was replaced by Pregothic script, another major book script (Figure 6.3). Letters lost some of their roundness and became pointier, while connections were forged between specific letter pairs. Moreover, the script developed an overall 'angular' and 'compressed' appearance. Throughout the twelfth century, this particular writing style spread across Europe, from Spain to Norway, and from Scotland to Italy. By the second quarter of the thirteenth century, however, it had developed into another script, called Littera textualis or Gothic textualis (Figure 6.4). This is the third major book script of medieval Europe, which flourished, with surprisingly little change, until the waning of the manuscript era, *c.* 1500.

Even though palaeographers instantaneously recognise which of these major book scripts is in front of them, the difference between the families is actually modest. Many letters remained unchanged, while others experienced only subtle changes. Additionally, fine differences in the execution of letters are observed *within* each family. There are also numerous regional variants of letter shapes within each family, so that

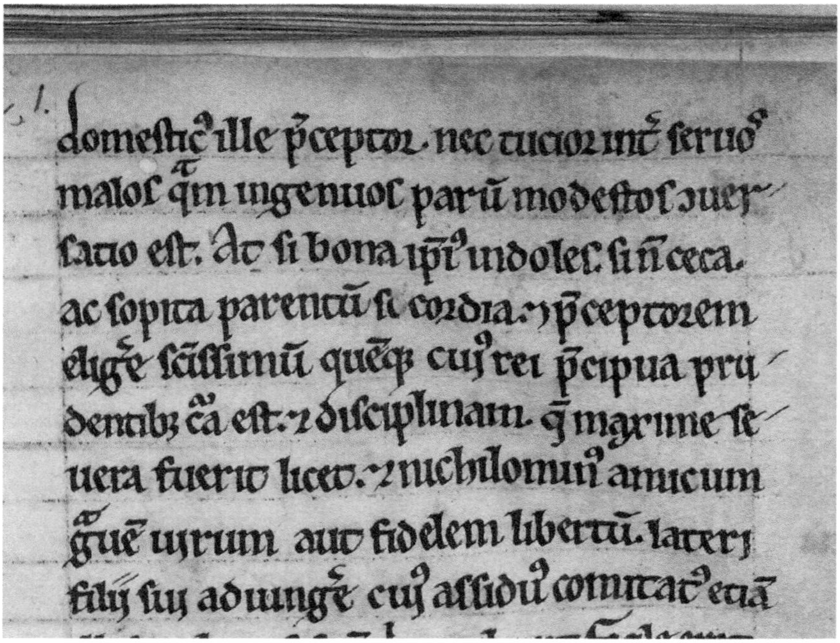

FIGURE 6.3 Pregothic script, *c.* 1150 (Leiden, Universiteitsbibliotheek, BPL 196, fol. 129v, detail).

the execution of Gothic script written in southern France is distinctly different from the look produced by scribes in Germany, or even by those in other parts of France. Moreover, scribes tended to add a personal flavour to their execution of a given script, bending conventions and producing unique variants. This all demonstrates just how much variation is present in the medieval script system, and how subtle it is.

Variation between, and within, script families can be articulated by defining how the scribe configured the individual strokes of letters. Medieval letters were rarely produced by a solitary stroke of the pen, which was the case with **i, I,** and long-**s**; that is, before the **i** became dotted, in the twelfth century, which required a second stroke. Usually, multiple pen movements were needed to produce a single letter, sometimes as many as three or four (**g** or **k**). How the scribe configured the strokes determined the ultimate shape of the letter, and the combination of all their choices determined the overall look of the scribe's

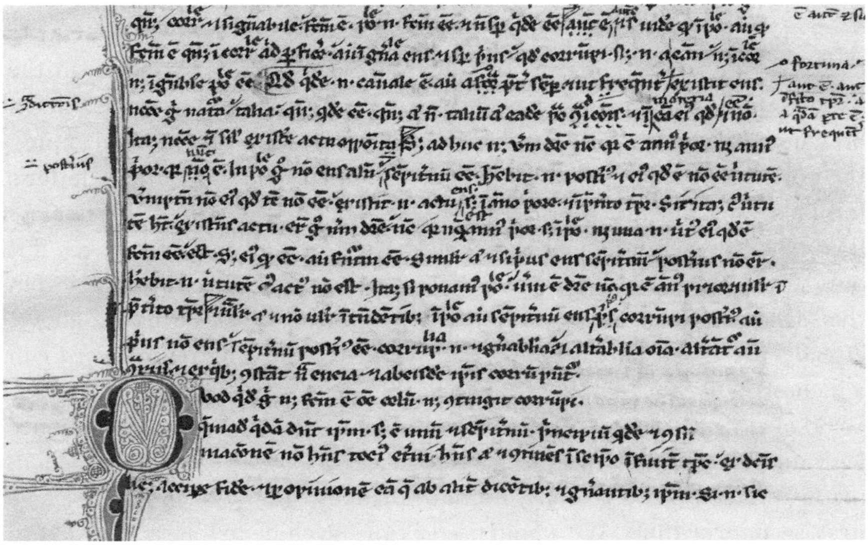

FIGURE 6.4 Littera textualis, 1250–1300 (Leiden, Universiteitsbibliotheek, BPL 64, fol. 147v, detail).

handwriting.[17] As mentioned earlier, the term 'duct' describes the process by which a letter was formed. In fact, two types of duct are usually distinguished. The first, 'basic duct', 'determine[s] the order and direction of the traces [...] required for the basic shapes of the letters in a particular script'.[18] Basic duct governs how the letters must be written for them to be recognised as belonging to a certain script family. Basic duct is an ideal, an image in the scribe's head, which makes Gothic script recognisable as Gothic script (although the name is modern in origin and has been assigned in retrospect by scholars). The second type, 'personal duct', 'determined the ways in which a scribe executed the traces of a basic duct'.[19] Personal duct defines how individual scribes shaped the letters of a given script family. It captures how scribes could vary, in subtle ways, the appearance of the letters within set conventions.

[17] Parkes, *Their Hands before Our Eyes*, p. 150 ('configuration').
[18] Parkes, *Their Hands before Our Eyes*, p. 151 ('ductus').
[19] Parkes, *Their Hands before Our Eyes*, p. 151 ('ductus').

Different Strokes for Different Folks

The reflections so far show how differences observed (or sensed) in the handwriting of two manuscripts are ultimately prompted by variations in how individual letters were configured by scribes. Crucially, determining the potential ways in which a stroke could be formed – defining a letter's full palaeographical breadth, as it were – makes it possible to articulate how two representations of a given letter are factually different, even if it only concerns subtle disparities. There are four variables in play within this system of stroke variation. The definition of 'basic duct', above, already presented two: scribes could vary, to some degree, the *order* in which strokes were written down as the letter was formed, while they could also vary the precise *direction* of the stroke. The first of these is not very useful here, because completed letters, as they are presented in manuscripts consulted today, do not normally show which stroke came first, second, or third. We would need a time machine to take a seat next to the medieval scribe to know for sure. Direction, however, can be observed and defined in clear terms, and is therefore a promising criterion.

Scribes often moved their pen in a particular direction as they executed a stroke, and to quantify the resulting stroke it can be helpful to super-impose its vanishing point on an imaginary clock face. Thus, scribes in the eleventh century tended to trace the 'tongue' stroke of the letter **e** towards three o'clock, producing a trace with a nearly perfect horizontal orientation (Figure 6.2, line 6, second word: 'atqu**e**'). During the twelfth century, however, the direction of this stroke was changed, and it came to point towards the number 2 on the clock's face (Figure 6.3, line 1, second word: 'ill**e**'). A more impactful change in direction is witnessed near the lower end or 'feet' of minims, the vertical strokes in letters such as **h, i, m,** and **n**. The scribe who copied the manuscript in Figure 6.3 lived in the middle of the twelfth century and he wrote the text in Pregothic script, a form of writing that came into use as the existing – ubiquitous – script family, Caroline minuscule, began to wane in Europe, as explained above. The general appearance of Pregothic script is influenced by a shift in the position of these feet, which came to turn to the right (Figure 6.3, line 1, first and fourth word: 'do**m**esticus', '**n**ec'), whereas in Caroline minuscule they either turned left or went straight down (Figure 6.2, line 1, second

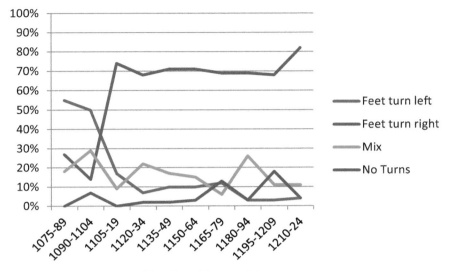

FIGURE 6.5 Direction of feet at minims.

word: 'neque').[20] Data derived from a sample of 353 manuscripts included in the *Catalogues des manuscrits datés*, which were gathered as part of my 'Turning Over a New Leaf' project (2010–2015) at Leiden University, show how this particular trend emerged and developed over time (Figure 6.5). The first manuscripts with feet consistently turning to the right appear near the end of the eleventh century, but it is in the first half of the twelfth century that this mannerism really takes off. Between 1100 and 1125 alone, the novel shaping of feet jumped by over 50 per cent, turning this shift in direction into a litmus test for determining whether a scribe has transitioned to writing Pregothic script.

Another way the medieval scribe could vary the execution of a stroke pertains to its shape. As Pregothic script emerged, scribes started to flatten round strokes, a process called 'angularity'.[21] For example,

---

[20] For a study of these feet, see E. Kwakkel, 'Biting, kissing and the treatment of feet: The transitional script of the long twelfth century', in E. Kwakkel, R. McKitterick, and R. Thomson (eds.), *Turning Over a New Leaf: Change and Development in the Medieval Manuscript*, Studies in Medieval and Renaissance Book Culture (Leiden: Leiden University Press, 2012), pp. 79–104.

[21] Kwakkel, 'Biting, kissing and the treatment of feet', p. 86.

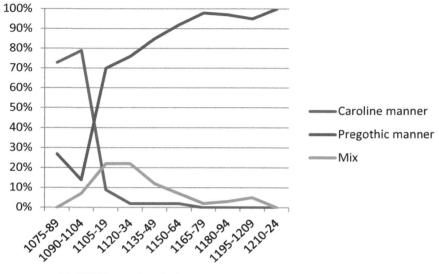

FIGURE 6.6 Angularity.

whereas in Caroline minuscule the two connecting strokes in the letter **m** formed round arches (Figure 6.2, line 4, fourth word: 'animis'), their appearance became flattened in Pregothic script: they began to resemble ski slopes rather than half circles (Figure 6.3, line 1, first and fourth word: 'domesticus', 'nec'). The same happened to other strokes with a round appearance, as in the case of the tops of the letters **c**, **e**, and **o**, and the 'bellies' of **b** and **d**. Data taken from the aforementioned 353 manuscripts show how angularity became an accepted mannerism in medieval Europe (Figure 6.6). As with the shifting direction of the feet on minims, the first quarter of the twelfth century is when flattened round strokes become popular, after a slow start in the late eleventh century. With so many letters affected in this way, twelfth-century script gained a pointy appearance; if one were to put a flat hand on top of these letters and push down hard, tiny pricks would now be felt all over. The quantified assessment, that the style of twelfth-century handwriting is pointy, presents a verdict similar to the impression that emerges from the script's aspect. The difference is that assessing the phenomenon through stroke configuration generates empirical evidence: when combined, stroke configuration and quantitative palaeography provide a way to quantify the expert's impressions.

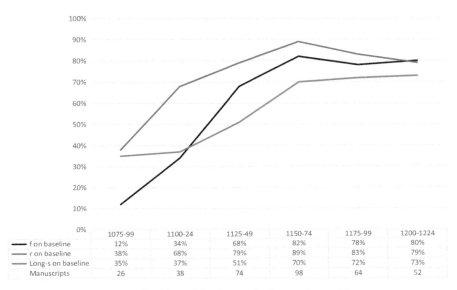

	1075-99	1100-24	1125-49	1150-74	1175-99	1200-1224
f on baseline	12%	34%	68%	82%	78%	80%
r on baseline	38%	68%	79%	89%	83%	79%
Long-s on baseline	35%	37%	51%	70%	72%	73%
Manuscripts	26	38	74	98	64	52

FIGURE 6.7 Position of the feet at the letters f, r, and long-s.

In addition to direction and shape, a stroke could also be varied in terms of its length. A particularly clear example is the length of letters with a pronounced vertical orientation, such as the **f**, **r**, and long-**s**. While in Caroline minuscule, the script used from *c.* 800 to *c.* 1075, these letters were extended below baseline (some minor extensions are visible in Figure 6.2, line 4, fourth and sixth word: 'animis', 'fingit'), in the last quarter of the eleventh century, as Pregothic script was emerging, a trend began by which the length of these three letters was reduced.[22] Ultimately, they came to be placed *on* baseline, rather than being extended below it (Figure 6.3, line 2, first word, '**m**alos'; line 4, third word, 'pa**r**entum'; line 8, fourth word, '**fi**delem'). The phenomenon took hold rapidly (Figure 6.7). It would ultimately become a standard feature of Littera textualis, the dominant European book script from *c.* 1225,

---

[22] E. Kwakkel, 'Book script', in E. Kwakkel and R. Thomson (eds.), *The European Book in the Twelfth Century* (Cambridge: Cambridge University Press, 2018), pp. 25–42 (p. 29).

albeit that the route towards full acceptance shows different speeds of progress for each of the letters. In fact, the trends caught in Figure 6.7 suggest that the **r** and long-**s** were already being placed on the line by a significant number of scribes before Pregothic script came into use. That is, by *c.* 1075 as many as a third of the manuscripts in the sample have the **r** and long-**s** consistently placed *on* the ruled line. In other words, while the shortened length would become a standard feature of Littera textualis, the trait itself was already present in older scripts. In fact, the feature occurs, to variable degrees, in all three major book scripts in Europe – Caroline minuscule, Pregothic script, and Littera textualis – making it more of a universal feature than a novelty exclusive to Pregothic script. This suggests a somewhat different look at script families: what varies is not the presence of this feature in Europe's major book scripts, but its popularity with scribes, who particularly favour reduced minims in certain eras.

Unlike the three variables introduced so far, the final way in which scribes could vary the execution of a letter concerns not so much how an individual stroke was configured, but how many strokes were used to produce a particular letter. While the tongue stroke at the letter **e** has a variable direction, as discussed, the very presence or absence of the tongue is also something that varied. Although it was present during the twelfth and thirteenth centuries, over the course of the fourteenth century, scribes started to drop the tongue at **e**. The result is that samples with the tongue still attached can commonly be placed in the first half of the fourteenth century, while those showing a tongue-less **e** are usually from the second half. An example from the realm of Pregothic script is the emergence of the 'dot' (commonly a small diagonal line) placed above the letter **i** (Figure 6.8). Here the new development is not a reduction in the number of strokes, as in **e**, but an increase. It is hard to know for sure why medieval scribes came to 'dot' the **i**, but, as the foot of the letter started to flick to the right, a trend discussed above, two letters **i** written consecutively started to resemble the shape of the letter **u**, now that the foot of the first **i** produced a closed connection with the second. The 'dot' was possibly a solution to this visual challenge, showing that some palaeographical change has a pragmatic motivation.

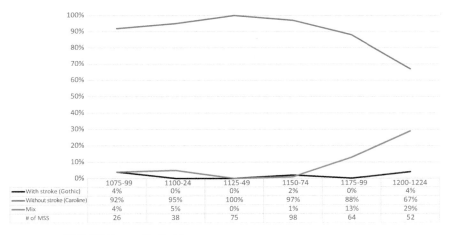

	1075-99	1100-24	1125-49	1150-74	1175-99	1200-1224
──With stroke (Gothic)	4%	0%	0%	2%	0%	4%
──Without stroke (Caroline)	92%	95%	100%	97%	88%	67%
──Mix	4%	5%	0%	1%	13%	29%
# of MSS	26	38	75	98	64	52

FIGURE 6.8 Dotted i.

Making Sense of Medieval Script

As these examples show, stroke variability provides an objective means to describe how medieval letters were configured. Moreover, due to its role in studying the process of writing in a quantitative manner by tracing the scribal mannerisms found in several hundreds of dated manuscripts, stroke variability helps reconstruct the history of a medieval script, from its birth, through its maturing stages, to its ultimate waning. Knowing when new configurations emerged and disappeared may prove important for determining the age of a manuscript. After all, assessing how strokes are configured in an *undated* manuscript, and contrasting these observations with the reconstructed chronologies of relevant mannerisms, may provide an indication of a manuscript's moment of production, even if only in broad strokes. For example, a scribe who refrains from using angularity, still extends the feet of **r** and long-**s** below the line, and produces a horizontal tongue stroke in **e**, most probably copied the manuscript before the twelfth century. In this way, stroke configuration may even come to support intuitive verdicts regarding the production date of a manuscript, or perhaps correct them. Most importantly, perhaps, the tables above show that quantitative studies of stroke configuration provide a peek under the bonnet and reveal some of the script's inner workings, particularly regarding the broader processes of change and development. The following highlights a few of the most striking observations that can be deduced from the data in Figures 6.5–6.8.

What the data in these figures show, for example, is how a medieval script is born. One may be inclined to think that a new script was adopted through a process whereby an increasing number of scribes embraced a newly introduced, fully fledged script – a view that may be enforced by the neat enumeration of a script's features in our handbooks and the introduction of script families in different chapters. Yet, data pertaining to Pregothic script suggests that a new script can appear on the scene intermittently and haphazardly. The tables suggest how, beginning in the eleventh century, individual deviations from Caroline minuscule were introduced, one or two at a time, usually in significant increments.[23] Just as features were being introduced at different moments, the novelties were not universally accepted at the same speed. Some of the early Pregothic traits were adopted in the late eleventh century, as reflected in, for example, the changing shape of round strokes (Figure 6.6) and the shortening of feet, placed on the line (Figure 6.7). Others, however, were introduced much later. The round version of the letter **s**, which came to supplement and ultimately replace the straight-**s**, arrived only in the second half of the twelfth century, as did the dotted **i** (Figure 6.8) and the phenomenon of 'biting', whereby contrary round letters placed adjacently would start to overlap, for example in **bo** and **de**.[24] In other words, there is no single beginning point of Pregothic script. Rather, there is a constellation of the beginnings of individual features, which would, as a group, ultimately come to define the script and become part of the new palaeographical norm for copying books.

Another example of a broader process that is revealed through quantitative data is the variable geographical adoption rate of new features.[25] To stay, for a moment, with the phenomenon whereby two contrary round letter forms start to overlap, by the first quarter of the thirteenth century almost 60 per cent of scribes trained in France fused the **de**

---

[23] Kwakkel, 'Book script' and 'Biting, kissing and the treatment of feet'.

[24] Kwakkel, 'Book script', p. 29, Figure 2.2 (round s) and 'Biting, kissing and the treatment of feet', p. 97, and Graph 4 at p. 207 (biting).

[25] Kwakkel, 'Book script', p. 30 and 'Biting, kissing and the treatment of feet'.

combination, as did some 35 per cent of their peers in England. At the same moment, however, the feature is still fully absent from manuscripts produced by scribes trained in Germany. The apparent reluctance to accept new features is seen more broadly among German-trained scribes. While, between 1125 and 1150, scribes in France had a broad palette of innovative mannerisms at their disposal, their counterparts in Germany included these in a very limited manner. This is why the aspect of German specimens feels much older than contemporary samples from elsewhere in Europe, which is ultimately something that has to be compensated for when a script is dated on the basis of its aspect. Moreover, this example shows how contemporaneous scribes in Europe wrote very different styles of Pregothic script. While, throughout the long twelfth century, scribes changed their writing habit significantly, they did not necessarily do so to the same extent or the same speed, highlighting the absence of a uniting principle or coordination behind the emergence of this particular new way of writing. It is all the more surprising, in this respect, that the different ways of writing Pregothic script ultimately converged and blended into a single norm, it appears. As scribes came to write more similarly across geographical regions in the second quarter of the thirteenth century, new norms appeared. This group of norms became the foundation of yet another new script: Littera textualis. This script came to incorporate, in a systematic manner, all the features that were slowly and haphazardly introduced over the course of the twelfth century.

This gaze into the inner workings of a script puts considerable stress on the reconstructed identity and genealogy of medieval script families, which is a particularly valuable impact of gathering hard data in a discipline built on soft evidence. Notably, in palaeographical handbooks the definitive state of Pregothic script (i.e., the full culmination of new features) is seen as the ultimate identity of the script, and not the intermittent appearance of individual novelties, which is more defining for Pregothic script as a whole, it turns out. Some features, moreover, never become fully accepted. Palaeographical handbooks regard the dotted i as a standard feature of Littera textualis, the dominant script used for the production of manuscripts from around *c.* 1225. However, as Figure 6.8 shows, at that time very few scribes consistently added the dot; the elevation of the mixed line shows that many did so only from time to time (see also Figure 6.3, where the only cases are observed in 'filii' in line 9). In fact, this particular change in the handwriting of scribes shows how a mixed occurrence of an older and newer presentation of a letter, a mix of two families, can become a standard in itself. While qualitative data taken from the corpus of 353 manuscripts do not tell us how this trend develops further into the thirteenth century, individual samples taken from dated manuscripts produced between 1225 and 1300 indicate that the mix of dotted and undotted **i** within a single manuscript would remain the standard attitude (note the dominance of the 'mix' line in Figure 6.8). As fascinating as this observation is, it challenges our system of mapping medieval script families. After all, Littera textualis, the later of the three major families, includes a presentation that was standard in Caroline minuscule (the undotted **i**), but it also includes the form that emerged in Pregothic script (the dotted **i**). Ultimately, this trend shows, then, that a new feature may not come to fully replace the existing presentation of a letter, but that the two sometimes coexist, blurring the clear-cut boundaries between scripts as defined in handbooks.

The presence of variation both between and within script families – to a degree that would have pleased Charles Darwin – prompts fascinating questions that invite us to rethink what a medieval script actually *is*. How many new mannerisms does a scribe need to embrace for the script to be labelled as new? Is a script family still the same if it has started to adopt new features? How many new features can an existing script absorb before it becomes something else? Is a new script born with the first emergence of new features, or when it has finished evolving and no more new features are introduced? Data based on quantitative research may not necessarily answer these questions, but that they prompt them is already a big win. Some of the most meaningful questions sparked by hard palaeographical evidence, however, are related to an issue that has been hiding in plain sight for much of the discussion so far: how did a new script feature take hold and become adopted by a growing number of scribes? How did a new style of writing move across the map of Europe and become established at a different location? These last questions highlight the true enigma of medieval letters.

The Enigma of Medieval Letters

Data related to Pregothic script show how European scribes slowly and unevenly adopted new letter forms, at least during the century and a half between *c.* 1075 and *c.* 1225. As a result, the new script did not advance as one entity, but it gradually gained shape as new features were introduced and took hold among a growing number of scribes. What the data do not reveal, however, is precisely how these new palaeographical mannerisms 'jumped' from one scribe to the next, resulting in the geographical spread that can now be observed. Did scribes feel inspired by the aesthetics of a new letter form they observed in the writing of their peers? Was it the gravitas or regulations coming from certain writing centres, such as monastic motherhouses, that sparked change? Or could it have been through education and the influence of individual writing teachers that new writing norms were transmitted?[26] And, more generally, just how quickly could a body of scribes in one location collectively adopt a new style of writing? One particular twelfth-century scriptorium, the Benedictine community in Rochester, Kent, provides a piece of this complex puzzle. Surviving manuscripts from the first quarter of the century indicate that Rochester scribes embraced new writing mannerisms in a surprisingly rapid and coordinated manner, very different from what is observed in the broader European sweep in the data discussed so far. This atypical scenario of change in writing practices is rooted in the historical context of the religious house.

The early-twelfth-century occupants of Rochester were all of Norman origin. During the abbacy of Gundulf of Bec, bishop between 1077 and his death in 1108, some 60 Benedictine Norman monks arrived. These individuals came both directly from Normandy and from the nearby Benedictine house of Christ Church, Canterbury, where a large contingent of Norman monks had settled in the wake of the Conquest.[27] These Norman roots explain why manuscripts produced in Rochester during the last quarter of

---

[26] For the influence of scribal training on script changes, see Kwakkel, 'Book script', pp. 38–39.

[27] For the Rochester case discussed here, see E. Kwakkel, 'Hidden in plain sight: Continental scribes in Rochester Cathedral Priory, 1075–1150', in E. Kwakkel (ed.), *Writing in Context: Insular Manuscript Culture, 500–1200*, Studies in Medieval and Renaissance Book Culture (Leiden: Leiden University Press, 2013), pp. 231–261.

the eleventh century are written in the typical Norman hand, representative of early Pregothic script. Feet had formed at minims, and these feet were all taking a right turn, as per Pregothic convention. Round strokes in **h, m, n, o, p,** and **r** had changed shape and received an angular appearance. Moreover, the general appearance of the script includes significant lateral compression and a 'tall' appearance: ascenders and descenders have an increased length. From around 1110, the Norman scribes in Rochester abandoned their native writing style, apparently quite suddenly. Manuscripts produced from that moment were written in what Neil Ker dubbed 'prickly script', a style characterised by sharp angles, flattened curves, pointy feet, and split ascenders (Figure 6.1). This Norman script on steroids is encountered in manuscripts written by Rochester scribes. A related, though distinctly different, version of prickly script was written in the Benedictine house of Christ Church, Canterbury, and it has been shown that this model was introduced in Rochester by monks who moved from Christ Church to Rochester in the late eleventh century.[28]

The story of the Rochester scribes suggests one possible resolution to the enigma of how a writing style moved across the map and became adopted by scribes in a different geographical location. Mannerisms, safely stored in the muscle memory of Canterbury monks, were transported to the new scriptorium at Rochester, where they were subsequently picked up by peers. It is possible that, sitting on the benches of the cold scriptorium, scribes observed new letter forms flowing out of the pens of the newcomers, which may have inspired them to imitate the prickly style in question. However, surviving manuscripts suggest there is more to the story. As already mentioned, the prickly style of Rochester is different from that in Christ Church.[29] In other words, what the newcomers brought with them was not simply imitated, but it was varied and used to create a new model. When one observes the surviving manuscripts produced in Rochester during the first quarter of the twelfth century, it also becomes clear that the majority of scribes, if not all, had adopted the new Rochester style of writing. This observation

---

[28] T. Webber, 'Script and manuscript production at Christ Church, Canterbury, after the Norman Conquest', in R. Eales and R. Sharpe (eds.), *Canterbury and the Norman Conquest: Churches, Saints and Scholars, 1066–1109* (London and Rio Grande: Hambledon Press, 1995), pp. 145–158.

[29] Webber, 'Script and manuscript production', p. 147.

suggests that in-house scribes were actively trained to write in the new script that was modelled on an 'imported' style from Christ Church, Canterbury.

The extent to which Rochester scribes in the first quarter of the twelfth century execute their letter forms in the same manner is truly remarkable. In fact, the individual duct of Rochester scribes of this period is so similar that it is sometimes impossible to distinguish between two scribes. This confirms that in-house scribes were trained, evidently in a rigorous manner, to write in the new style. Further, the similarity of individual hands also evidences that scribes could alter their existing mannerisms even at a later age, decades after they had settled into a specific way of writing (in this case the Norman style of Pregothic script, which was abandoned for the new prickly script). While the Rochester case is perhaps exceptional in a broader European context, since such quick and consistent changes among a large group of scribes seems quite unusual, it helps to chip away at the enigma of medieval script, specifically as to how new writing features were disseminated. New letter forms travelled and became models for new users. In Rochester, prickly letters were transported to a new home and, upon arrival, became accepted by a large group of individuals, although in Rochester the letterforms were then employed in a slightly adjusted state. Notably, although in-house scribes changed their traditional writing style rather quickly and consistently, they still reverted, from time to time, to their old habits. It is when they tested their pens that the 'suppressed' older script mannerisms appear for a short moment: the numerous pen trials on surviving flyleaves in Oxford, Bodleian Library, Bodley MSS 340 and 342 reveal, through their precise letter forms, the continental origins of Rochester scribes.[30] Apart from these instances, their attitude towards the acceptance of a new writing style is remarkably different from the broader European sample of scribes captured in the tables above: in Rochester, change came not with intervals and at a slow speed, but suddenly and completely. Considering that medieval scribes were apparently able to bring about consistent change in a very short amount of time, it remains all the more enigmatic that European script in the twelfth century developed in such a slow and haphazard manner.

[30] For this case, see Kwakkel, 'Hidden in plain sight'.

7 Eruptions, Emissions, and Enigmas: From Fuming Volcanic Vents to Mass Extinction Events

TAMSIN A. MATHER

Introduction

Volcanoes are enigmatic in many different ways. They can be huge threats to lives and livelihoods, but are also key to many important natural resources, such as geothermal power and ore deposits, and are at the centre of communities in various ways. They have almost certainly played a role in maintaining the habitability of our planet, but they have also been implicated in some of the biggest upheavals in life's evolution. I will touch upon these themes in this chapter. I also want to give you a sense of my personal experience as a volcanologist – why I feel I am peering into the mysteries of the deep Earth and deep time when gazing into the vent of an active volcano.

Volcanoes and the Enigmatic Earth beneath Our Feet

Let me start with a photo (Figure 7.1). This was taken on the crater rim of Masaya Volcano in Nicaragua at the end of my first term as a PhD student in Cambridge. Many of you might find this photo vaguely scary because you can see a massive crater yawning behind me with an ominous haze wafting out of it. You will be pleased to see that I am paying attention to health and safety: helmet on and gas mask to hand. The different pieces of equipment in the photo are to collect the volcanic gases and particles in the haze behind me to take back to the lab for analysis.

This was my first campaign gathering samples for my thesis and, in fact, my first time doing science outdoors. It was also my first visit to Latin America, and we had a troublesome journey. The airline had lost three of our bags, containing most of the equipment and all my

FIGURE 7.1 Tamsin A. Mather at Masaya volcano, Nicaragua, December 2001. Photo: Clive Oppenheimer.

colleague's clothes. This meant that it was fully dark when we tried to find the hotel. With no access to smartphones and instant Google Maps (it was 2001), we missed a vital turning, got lost, and eventually arrived extremely late. The next morning, I woke up with the sunlight creeping through the thin wooden slats of my room and the sound of termites eating the walls from the inside. When I opened the door, I was greeted by a view of the fuming Masaya volcano seemingly only a stone's throw away through the jungle (Figure 7.2). I had never been anywhere quite like it!

'Summit of Masaya volcano' might sound a little too grand. It is, after all, a low-altitude volcano, only just over 600 metres above sea level. This ease of access and the long-term gas emissions from Masaya's vent make it an excellent place to study volcanic processes. I have heard colleagues call it a drive-through volcano: you can drive your equipment

FIGURE 7.2 View towards Masaya from the Hotel Volcan Masaya in 2001. Photo: Clive Oppenheimer.

right up to the crater and run your experiments. (I must admit that this description makes me wince, with its connotations of Starbucks and McDonald's so discordant with the emotional space Masaya inhabits for me.)

This was my first volcanic expedition, and I did not know what to expect. Figure 7.3 shows what I was able to see as I peered into the crater. You can see the gas and particle plume pumping out of the two vents that were present in 2001, but there was no glow from the magma in the daytime and only a faint glow at night. When the wind dropped, you could hear the liquid rock sloshing around, roaring in fact, but the main clue to its presence was the persistent smell of the gas transported with it, rising from the depth.

Over the years, a collapse of the crater floor, rock falls, and perhaps some small rise in the magma level have revealed more of what's going on beneath Masaya's surface. Last time I went, in 2017, we were able to stare

FIGURE 7.3 View into Masaya's active crater in 2001. Photo: Clive Oppenheimer.

pretty much directly into the top of the magma column. Figure 7.4 does not really do it justice: in person the magma was a roiling current of red-hot energy sloshing around, almost like a hellish ocean blowhole with gas whooshing out, accompanied by a constant pulsating roar. Though more was visible, Masaya was still very enigmatic. These were only the final seconds of a magmatic journey that started some 100 kilometres below the surface, and we get little sense of the magma's path once it loses its gas and sinks back into the depths. This is especially enigmatic because comparatively little magma has erupted over Masaya's history.

Peering into this 'mouth of hell' (as sixteenth-century Spanish Conquistadors described it) felt like tracing clues about the inner workings of our planet. These workings are very enigmatic, and, without the well-ordered, controlled circumstances of a chemistry lab, we are in many ways at nature's whim in terms of the evidence that we can gather. One of the things that really fascinates me about Earth Sciences is the sense of

FIGURE 7.4 View into Masaya's active crater in 2017. Photo: Tamsin A. Mather.

mystery, in some ways vertigo, that I get when standing on the skin of our planet, trying to understand what is going on beneath our feet. For example, imagine the core some 3,000 kilometres below. Every day, we feel the benefits of this slowly freezing but partly pulsating metal-rich mass: it makes our compasses swing true and shields us from harmful solar wind. Despite its influence, we cannot look at it or take samples. Through volcanoes, however, we *can* get hints about our planet's interior workings. This is part of what has driven me since my PhD.

There is another side to these enigmas. Questions about the nature of the Earth's core can seem remote from human experience, but volcanoes are a different story. Fatal eruptions like those on White Island in New Zealand and Taal Volcano in the Philippines (which occurred just before I gave the lecture on which this essay is based) remind us that our lack of understanding of the enigmas beneath our feet can have very real consequences for those in the wrong place at the wrong time.

The Enigmatic Link between Volcanoes and Global Change Deep in the Geological Record

As important as understanding eruption precursors is, this chapter will not focus on volcanoes as hazards. Instead, I will discuss a volcano-related enigma that, for me, relates to the Darwin lecture series. On 20 February 2004, in the final year of my PhD, I came to a lecture by Vincent Courtillot from Paris that was part of the Evidence series.[1] (I must admit that I am slightly concerned that lectures on volcanoes seem to have evolved from being part of the Evidence series to inclusion in the Enigmas one!)

This was one of the first times that I saw Vincent's correlation between the ages of flood basalts and mass extinctions (Figure 7.5). The figure shows the age of extinction events and other major environmental shifts as recorded in the geological record on the y-axis, against the age of continental flood basalts or oceanic plateaus (collectively referred to as 'large igneous provinces', as discussed in the next section) on the x-axis, both in millions of years before the present. The plot presented by Vincent was a summary of improvements in the dating of the sediments and basalts, obtained by himself and others. I vividly remember being intrigued by this figure, and, although my own work did not immediately turn to it, I think it registered somewhere. It is no surprise that the sinuous trail of science has brought me back to it.

The timescales shown in this plot are quite dizzying from a human perspective. When I started my PhD, I attended a lecture where the speaker put his arm out and asked us to guess how long human beings had existed, if his shoulder represented the 4.5-billion-year age of the Earth and his fingertips the present day. I thought maybe we would get to a couple of knuckles, but actually, if you take a nail file and draw it across your fingers once, that is roughly the proportion of time that human beings have been around. It made quite an impression on me; it is always worth remembering that we are just the nail dust of planet Earth.

---

[1] V. E. Courtillot, 'Evidence for catastrophes in the evolutions of life and earth', in A. Bell, J. Swenson-Wright, and K. Tybjerg (eds.), *Evidence* (Cambridge: Cambridge University Press, 2008), pp. 48–72.

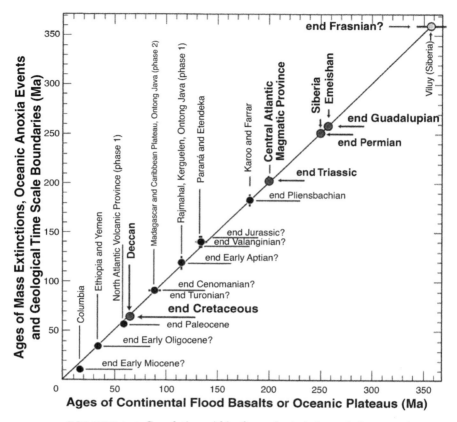

FIGURE 7.5 Correlation within the geological timescale between the ages of large igneous provinces (continental flood basalts and oceanic plateaus) and those of mass extinctions and oceanic anoxia events (in millions of years, Ma). Source: V. E. Courtillot and P. R. Renne, 'On the ages of flood basalt events', *C. R. Geoscience*, 335 (2003), 113–140, Fig. 1. © 2003 Académie des sciences, published by Elsevier Masson SAS. All rights reserved.

Indeed, getting your head around these baffling timescales is one of the excitements of Earth Sciences.

Vincent's starting place in his Darwin lecture was the end-Cretaceous mass extinction about 66 million years ago. You may well have heard more about this extinction than of the others: it is the most recent, and it marked the demise of the dinosaurs. This event also gets a lot of press because there has been an ongoing debate about whether it was the

Chicxulub impactor hitting Earth that killed the dinosaurs, or whether it was the large igneous province volcanism of the Deccan Traps.

Large Igneous Province Volcanism

Large igneous provinces are not your typical type of volcanism. Figure 7.6 shows an example of the scenery they produced in the most recent flood basalt on our planet: the Columbia River flood basalt that erupted about 17 million years ago. You can see stack upon stack, layer upon layer of lava flow over a very, very wide area creating this classic

FIGURE 7.6 Large igneous provinces: a stack of lava flows, typical of the 'trap' or stepped topography characteristic of the flood basalt lavas that make up large igneous provinces, exposed along the Grande Ronde River in Washington State (USA). Source: S. Self, A. Schmidt, and T. A. Mather, 'Emplacement characteristics, time scales, and volcanic gas release rates of continental flood basalt eruptions on Earth', in G. Keller and A. C. Kerr (eds.), *Volcanism, Impacts and Mass Extinctions: Causes and Effects, Geological Society of America Special Paper* 505 (2014), pp. 319–337, doi:10.1130/2014.2505(16). © 2014, Geological Society of America. Licence: CC-BY.

trap topography. To give you an idea of the area these provinces cover, Figure 7.7 shows the Deccan Traps that erupted about 66 million years ago (end Cretaceous period), set against the scale of the subcontinent of India. The Deccan is made up of about a million cubic kilometres of lava which erupted over several million years, but with much activity being concentrated within about one million years.

A million cubic kilometres is a difficult number to get your head around, so I did a few sums. Just a single cubic kilometre of lava would cover the whole of Cambridge in lava about 25 metres deep or, alternatively, Greater London in lava about 60 centimetres deep. Apparently, however, the usual unit of catastrophic area is Wales, so I should report that it would cover the entirety of Wales in about 5 centimetres of lava.

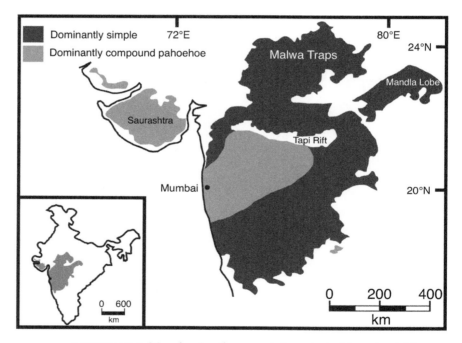

FIGURE 7.7 Map showing the present-day extent of the subaerial Deccan Traps, India (inset shows their scale compared with India as a whole). Source: T. A. Mather and D. M. Pyle, 'Volcanic emissions: short-term perturbations, long-term consequences and global environmental change', in A. Schmidt, K. Fristad, and L. T. Elkins-Tanton (eds.), *Volcanism and Global Environmental Change* (Cambridge: Cambridge University Press, 2015), pp. 208–227, Fig. 14.1(d), doi:10.1017/CBO9781107415683.018.

As mentioned, the Deccan traps released about a million of these cubic kilometres, which would be Wales about 50 kilometres deep in lava. This outpouring of lava also pumped enormous quantities of gas out into the atmosphere, for example about a million million tonnes of sulphur dioxide. Although we can average these eruptions of lava and gas emissions over the duration of the province's emplacement, in reality the activity is likely to have been highly pulsed, with high-intensity bursts punctuating periods of pause, the durations of which are hard to assess as they did not leave much of a trace in the rock record.

I have mentioned the Columbia River and Deccan Traps as examples of large igneous provinces, but actually the surface of our planet is peppered with their remnants. Figure 7.8 illustrates their distribution. Blue areas indicate large igneous provinces that erupted largely under the ocean; red areas erupted subaerially, that is, onto the continents; yellow areas are more silica-rich, rather than basaltic, and I will not discuss them here. There are two areas I will mention later: the Central Atlantic Magmatic Province, torn apart by tectonic forces, now spread between the

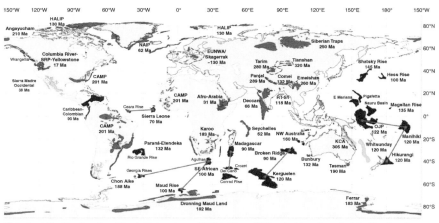

FIGURE 7.8 Large igneous provinces of the world, including continental flood basalt provinces, oceanic plateaus, and silicic large igneous provinces (LIPs). Source: S. E. Bryan and L. Ferrari, 'Large igneous provinces and silicic large igneous provinces: Progress in our understanding over the past 25 years', *Geological Society of America Bulletin*, 125 (2013), 1053–1078, Fig. 1, doi:10.1130/B30820.1.

continents of North America, South America, and Africa, and the enormous area of the Siberian Traps in present-day northern Russia. I was once lucky to get clear views of these on a flight from London to Tokyo and spent what felt like hours gazing at this amazing trap topography – the enormous area of basalt that was put out onto the surface of our planet.

While Figure 7.8 illustrates that large igneous province volcanism is not completely rare in terms of the geological record, we should note that we have not experienced the eruption of a large igneous province during the history of our species (the Columbia River is the most recent one, 17 million years ago). Hence, we must knit together clues from a number of sources to understand their impacts on our planet's environment.

Volcanism and Its Impact on the Earth's Environment

As a volcanologist, I try to use active volcanoes to learn what we can about their influence on the Earth's environment. A key motivation for this is to understand more about the impact of extraordinary periods of volcanism deep in the Earth's past. We can categorise volcanic activity in a number of different ways. Certain types of activity tend to make the news, for example when people are directly affected or when it generates spectacular photos. But volcanic activity is present every day at lots of locations around the planet, much of it relatively 'quiet', like the release of gas from a lava vent at Masaya. Sometimes, there is no magma at the surface at all, with volcanic gas seeping out of cracks in the earth (fumaroles), or more covertly still as soil gas. This is in stark contrast with rarer, explosive eruptions like Mount St Helens in 1980 (Figure 7.9), where you can see an enormous Plinian[2] column punching gas up into the stratosphere and scattering ash on a continental scale.

Also, as mentioned earlier, there are types of volcanism that we have not experienced in historic times. This includes large igneous provinces, but there are also very large eruptions, sometimes called super eruptions, that release over 1,000 cubic kilometres of magma (Mount St Helens erupted about 0.25 cubic kilometres). One of the last super eruptions took

---

[2] This is named after Pliny the Younger's description of the eruption of Vesuvius in AD 79.

FIGURE 7.9 Plinian column from 18 May 1980 eruption of Mount St Helens, USA. Aerial view from southwest with Mount Adams in the background. Photo: Robert Krimmel, US Geological Survey.

place at the Toba caldera in Sumatra, Indonesia, about 75,000 years ago. The most recent known super eruption was in the Taupō Volcanic Zone, New Zealand, about 26,500 years ago.

Though I also use tools like satellite measurements, often I need to make measurements inside actual volcanic emission plumes. For obvious reasons, I mainly make these at places like Masaya, at the less violent end of the spectrum. The challenge is to extrapolate the type of environmental impact that we get from these volcanoes to the possible fallout from things like large igneous provinces, which could cause ecological disturbance on the scale of mass extinctions.

To think about how to do this, let us first take a step back and consider what volcanoes release into our environment. I have mentioned lava flows with reference to trap topography. Though less characteristic of basaltic large igneous province volcanism, there are also other types of flow, for example pyroclastic flows – hot clouds that race down the volcanic landscape.[3] These and other immediate volcanic hazards are, while potentially lethal, relatively localised in terms of their effects. In contrast, wider-scale environmental impacts are largely mediated through the atmosphere and oceans. I have already alluded to ash and sulphur dioxide from volcanic activity, but these are just some of the ingredients of a whole cocktail of different gases and particles that volcanoes pump out.

Water is a big part of this cocktail, unsurprisingly given the ubiquity of it on our planet. However, our atmosphere already has a lot of water in it, which varies considerably through cycles of evaporation, condensation, and precipitation. Therefore, despite the large quantities, the water that comes out of volcanoes does not usually change the environment of our planet.

Second on the list is often carbon dioxide, CO_2. Its potentially long atmospheric lifetime and its impact on Earth's climate and oceans are sadly well documented. On the individual eruption scale, however, volcanic CO_2 emissions are insignificant compared with other sources. Although it is challenging to make very complete measurements, even our best maximum estimate of carbon dioxide emitted by the world's volcanoes today is about 100 times less than that from human activities.[4] In fact, there is evidence that CO_2 from 'normal' levels of volcanic activity, like those we see today, is part of a longer-term global carbon cycle that stabilises Earth's environment. However, the million years or so that a large igneous province is active sees increased global rates of volcanism with the potential to perturb the planet's carbon cycle over geologically significant timescales.

After water and carbon dioxide, we get into the less abundant but more reactive gases in the atmosphere. The sulphur gases – sulphur dioxide and hydrogen sulphide – are often significant in their abundance. In fact, these are the gases you might notice most in volcanic areas because of their distinctive smell. If you smell rotten eggs, that is hydrogen sulphide; something more like burnt matches signals sulphur dioxide. I was describing these smells in some detail on a programme for the BBC

---

[3] It was pyroclastic flows from Vesuvius that buried Pompeii in AD 79.
[4] C. A. Suarez, M. Edmonds, and A. P. Jones, 'Earth catastrophes and their impact on the carbon cycle', *Elements: An International Magazine of Mineralogy, Geochemistry, and Petrology*, 15(5) (2019), 301–306.

World Service a few years back and the presenter turned to me and said 'Are you a connoisseur of volcanic gases?' Maybe you could say that, but, unfortunately, I have never been able to calibrate my nostrils. I would love to avoid carrying heavy equipment up the volcano, and just stand downwind and take a nice deep breath to discern the sulphur dioxide/ hydrogen sulphide ratio, but we are stuck with electronics for now. In any case, sulphur gases have important atmospheric consequences that I will come to shortly.

More minor gases emitted by volcanoes include acidic gases like the hydrogen halides (for example hydrogen chloride, hydrogen bromide, and hydrogen fluoride), and some low levels of metallic vapours like mercury.

The ash and gas parts of a volcanic plume are intermingled with fine droplets called aerosols. These aerosols are frequently water- or sul- phuric acid-based and look like a fine haze, which can contain metals like lead, copper, and even gold. The acidity of the droplets might also leach other elements from the ash of the plume. Consequently, volcanoes throw out much of the periodic table into our environment, and these different elements can have various impacts as they move through the atmosphere, biosphere, and oceans.

Different volcanoes release these components in different balances – in fact, a single volcano can simultaneously release different balances as well, which is illustrated in Figure 7.10. We see the 2001 eruption of Mount Etna, with three types of volcanic plume, each with a different colour signalling a different chemistry. Furthest away is the darker, ash-rich plume coming from the new cinder cone forming from the flank eruption south-east of the summit. Closest to the camera is a white steam and aerosol-rich plume wafting out of the north-east crater, much like it does outside of eruptive periods. In between is the plume coming from the then single-summit south-east crater (which has now changed). It resembles in some way a colour palette mix between the other two plumes. Clearly, even at the same volcano at the same time, different emission processes lead to different chemical cocktails being put out into the atmosphere.

Though we learn a lot by looking at volcanic plumes with the naked eye, modern technology provides a whole range of other tools. In par- ticular, satellite-based observations can give us crucial new insights. An important eruption for volcanologists was that of Mount Pinatubo in the

FIGURE 7.10 Aerial view looking south at the summit of Mount Etna, Italy on 1 August 2001. Photo: Clive Oppenheimer.

Philippines on 15 June 1991. This was a Plinian type of eruption that put out about 5 cubic kilometres of magma and around 20 million tonnes of sulphur dioxide, with an eruption column that punched up to 35 kilometres into the atmosphere. It can be thought of as a once-in-a-century event: neither super common nor super rare. One of the reasons why this is an important eruption is that it is the largest that has happened in the satellite era.

The height of a volcanic plume is an important indicator of the range of impacts it could potentially have. Roughly speaking, the higher an eruption gets, the further its potential reach, but whether the material stays in the lowest part of the atmosphere (troposphere) or gets into the next level (stratosphere) is an important distinction. The weather and convection of the troposphere strip out reactive plume components much more quickly than the drier, more stable stratosphere, so that reactive gases like sulphur dioxide can hang around much longer higher up.

After the Pinatubo eruption, we were able to study the enigmatic effects of volcanoes high up in our atmosphere in a way that we had not been able to do before. Figure 7.11 shows photographs taken from a space shuttle. Comparing the two images, you can see a strange optical effect up in the stratosphere after the eruption, caused by the products of the volcano. As well as observing these effects with the naked eye, scientists used satellites to track what happened with the products of Pinatubo on a global scale. Figure 7.12 shows satellite measurements of optical depth (a measure of haziness) in the stratosphere. The figure shows a clear, clean stratosphere before eruption, followed by haziness spreading rapidly around the equator from the eruption site in the Philippines, then a slower spreading towards the poles, so that by August and September 1991 most of the global stratosphere was hazy. This haze persisted: the final image shows a misty stratosphere into 1994, well over two years after the eruption.

But what caused this? We know from everyday experience that fine particles in the atmosphere make it hazy. If you spray an air freshener, you see a haze that rapidly becomes transparent. More poetically, imagine water droplets on a spider's web, caused by an early Fenland mist, slowly being burnt off by the sun and dissolving into a haze. Translating this to Figure 7.12, the higher the optical depth, the more fine particles are present in the atmosphere. These are not ash particles; they would not persist long enough. What we are seeing is something more like the haze wafting out of Masaya's mouth in Figures 7.3 and 7.4: a fine mist of sulphuric acid droplets caused by the reaction of the volcanic sulphur dioxide with other gases in the atmosphere.

Planetary aerosol veils are not just of academic interest, they have global consequences. For example, observations have shown that temperatures on the Earth's surface were lower on average for several years after the Pinatubo eruption.[5] This is caused by the interaction of the particles with the Sun's radiant energy. Again, we know from everyday experience that you can see further through a clear atmosphere than through a hazy one, because light gets scattered by particles in a mist.

---

[5] A. Robock, 'Volcanic eruptions and climate', *Reviews of Geophysics*, 38(2) (2000), 191–219.

August 30, 1984

August 8, 1991

FIGURE 7.11 Two photos taken by astronauts from orbit showing a vertical cross section through the Earth's atmosphere at sunset before (upper panel) and after (lower panel) the Mount Pinatubo eruption. Image courtesy of NASA's Earth Observatory. Source: https://earthobservatory.nasa.gov/.

SAGE II 1020 nm Optical Depth

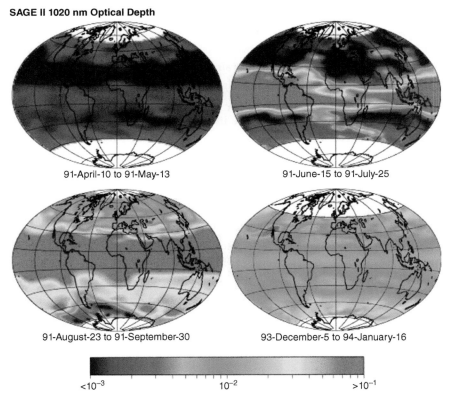

FIGURE 7.12 Images acquired by the Stratospheric Aerosol and Gas Experiment II (SAGE II) flying aboard NASA's Earth Radiation Budget Satellite (ERBS). Image courtesy of NASA's Earth Observatory. Source: https://earthobservatory.nasa.gov/ (see text for explanation).

After an eruption, the small sulphuric acid particles up in the stratosphere scatter the Sun's radiation and reflect more of its energy back into space. If you had been an astronaut or an alien sitting on Mars or on the moon, Earth would have looked a tiny little bit brighter after the Pinatubo eruption, while on Earth, less energy reaching the surface meant cooler temperatures. All in all, the scale of the eruption of Pinatubo and the global datasets available gave scientists new insights into how and why volcanoes influence the global climate.

Returning to the impact of large igneous provinces, an important question is what style of eruption they have. This is difficult to investigate because deposits from large igneous provinces are very old, and the ones left behind do not resemble the deposits of big Plinian eruption columns like the Pinatubo case (see Figure 7.6). Instead, the activity was probably of a more persistent type, pumping out lots of gas into the lower atmosphere over long periods of time, while only sometimes generating higher eruption plumes during periods of more explosive behaviour. This highlights the importance of studying both lower-altitude, persistent volcanic emissions, like those of Masaya, and high-altitude, short-lived injections, like Pinatubo.

From Masaya to Large Igneous Provinces

As mentioned, Masaya is a relative footnote on the Nicaraguan landscape in terms of altitude, and there is a lot of land at similar altitude that is regularly fumigated by its emissions. This provides a scientific opportunity to investigate the chemistry that occurs in low-altitude plumes. At the same time, communities live in this plume-bathed environment, so there is also a very human angle on this work.

The closest downwind community to Masaya is El Panama, a settlement of cinder block houses along a dirt track on the rim of the old caldera, only about 2 kilometres from the active vent. Masaya was not always releasing gas: historical records show that its degassing switches on and off, and the latest episode started in the early 1990s. Some people moved into the community before this time, and they have seen their lives change in very noticeable ways.

This is immediately apparent from the vegetation. Although this part of Nicaragua is characterised by lush growth (Figure 7.2), the fumigated environment downwind of the volcano is more like bleached, scrubby grassland. Coffee (Nicaragua's main cash crop) does not react well to the acidic gases, nor do typical crops like beans and maize. People had to adapt their agriculture to pineapples and dragon fruit, which grow well in this environment. The community is very poor, and it has problems such as lack of access to running water, but the volcanic gases add additional challenges – they can cause breathing difficulties, tiredness, water contamination, and the corrosion of metal building materials.

A project I was involved with made a short film about the community, which included a snippet of the roiling magma down the vent at Masaya.

Last time I was there, in 2017, with the help of the local agency INETER we set up a big screen and played the film in the local school. The audience loved it, watched it several times over, and would probably have watched it all evening if we had been able to stay. Because of the admission cost, many had never visited the Masaya National Park and had never looked into the active vent. The lava in the film gave them an image of the source of the gas that affects them every single day of their lives.

To better understand large igneous province-type eruptions, we need to scale up from the activity at Masaya and study close analogues in historical times. The closest parallel is probably the Icelandic 1783–1784 Laki eruption that poured out about 14 cubic kilometres of lava and led to famine and the deaths of an estimated quarter of the population. More recently in Iceland, there was the smaller 2014–2015 Holuhraun eruption, which totalled about a cubic kilometre of lava (Figures 7.13 and 7.14). These eruptions, characterised by sporadic,

FIGURE 7.13 Fire-fountaining (estimated heights of 70–100 metres) at the Holuhraun eruption (Iceland) in September 2014. Photo: Evgenia Ilyinskaya.

FIGURE 7.14 Bird's eye view of the lava flow and emissions from the fissure eruption at Holuhraun on 21 January 2015. Photograph taken from an Icelandic Coast Guard helicopter. Photo: Anja Schmidt.

violent fire fountain episodes and long-lived, less explosive lava activity, are probably most similar to large igneous provinces. Combining ground-based measurements, satellite data, and atmospheric modelling following Holuhraun, colleagues and I were able to learn new things about the scale and range of the atmospheric and environmental impact of this type of eruption.[6]

---

[6] T. A. Mather and A. Schmidt, 'Environmental effects of volcanic volatile fluxes from subaerial large igneous provinces', in R. E. Ernst, A. J. Dickson, and A. Bekker (eds.), *Large Igneous Provinces: A Driver of Global Environmental and Biotic Changes*, AGU Geophysical Monograph (New York, NY: Wiley, 2021), pp. 103–116.

The Complex Relationship between Large Igneous Provinces and Global Change in the Geological Record

Although there is much we can learn from present-day volcanism, scaling up in size and duration to the impact of large igneous province volcanism remains a challenge. It is also difficult to understand the differences between the present-day planet and the world tens to hundreds of millions of years ago. To solve this puzzle, we need to fully interrogate the geological record of global environmental change, and to determine its relationship with large-scale volcanism.

Returning to the plot that Vincent showed during his Darwin lecture in 2004 (Figure 7.5), there is quite a bit of detail beyond the mass extinctions that I do not remember him exploring at the time. A big five of mass extinction events are often referred to in the geological record: End Ordovician (450–440 million years ago), Late Devonian (375–360 million years ago), Permian–Triassic (about 252 million years ago), Triassic–Jurassic (about 201.3 million years ago), and the Cretaceous–Paleogene extinction event mentioned before (about 66 million years ago). Four of these are shown in Figure 7.5 ('end Frasnian' relates to the Late Devonian). The rest of the labels under the correlation line refer to minor mass extinction events, oceanic anoxic events (intervals when large parts of an ocean were depleted of oxygen), or other environmental events indicated by things like abrupt changes in the carbon isotopes recorded in Earth's sediments. The plot correlates these events with the timing of large igneous provinces, but the phenomena are so diverse that it is impossible to deduce a one-to-one relationship. There are other significant things that could be noted. The end Jurassic, for example, is not accompanied by a known large igneous province, and there are considerable uncertainties about the apparently limited environmental impacts of provinces such as the Ethiopian and Columbia River. I believe these variations give a richness that may allow us to understand more about the Earth's response to different imbalances and forcings.

Another interesting observation relates to the End Ordovician mass extinction (the earliest member of the 'big five' and not covered by the time period shown in Figure 7.5): we have yet to find evidence of a contemporaneous large igneous province. Is it possible that the volcanic rocks have weathered away, or have been removed by subduction, crushed by tectonic forces, or overprinted by another large igneous province? Or was there no large igneous province at all, so that the correlation with the 'big five' is incomplete?

Let us step back for a moment and consider some of the challenges and complexities that scientists face when making these temporal associations in the geological record. Mass extinction events are identified through the fossil record by measuring total species diversity during a particular period. Darwin himself was not a huge fan of the fossil record: he likened it to trying to read a story from a few pages torn out of a book. Nevertheless, in Darwin's day, the idea of looking at species diversity was already being discussed. John Phillips produced one of the first plots of species diversity over the geological period, recorded in British rocks in the mid 1800s. In his 1860 book *Life on the Earth: Its Origin and Succession*, his Figure 4 identifies both the end Permian and the end Cretaceous mass extinctions. On the opposite page, it says 'we have found by this mode of enquiry that the abundance of the forms of life in the sea has been very unequal at different periods and that race has followed race so as to match the words of the poet', and then it goes into Latin from Lucretius. I believe it translates as 'some people wax, others wane, in a brief interval the generations replace each other and like the racers transmit from hand-to-hand the torch of life'. We do not write scientific papers like we used to!

Phillips spent some of his career in Oxford. There is a bust of him in the Oxford University Natural History Museum, which is just around the corner from my office. In terms of Phillips's personal handing of the torch of life from generation to generation, he had what was described by one of my sources as a rather 'Oxford end'. He had a very fine dinner at All Souls College, in very illustrious company, then tripped over a carpet and fell down the stairs and died. I am not sure whether this is quite what Lucretius had in mind.

Returning to my point about the challenges in drawing temporal associations between change events in the geological record and large igneous provinces, I will zoom in on the end-Triassic mass extinction, about 201 million years ago. The end-Triassic world looked very different from that of today, with the continents largely fused together into a

FIGURE 7.15 Reconstruction of the end-Triassic world, with the modern continents overlain. CAMP, Central Atlantic Magmatic Province. Source: L. M. E. Percival, M. Ruhl, S. P. Hesselbo, H. C. Jenkyns, T. A. Mather et al., 'Mercury evidence for pulsed volcanism during the end-Triassic mass extinction', *Proceedings of the National Academy of Sciences*, 114 (2017), 7929–7934 (at 7930, Fig. 1), doi:10.1073/pnas.1705378114.

supercontinent (Figure 7.15). The Central Atlantic Magmatic Province erupted in the heart of this supercontinent, and its rocky remains have been torn apart onto the three separate present-day continents of Africa, South America, and North America. In the end-Triassic mass extinction, about 70 per cent of all species were lost. Hard-hit ones included the eel-like conodonts in the oceans, and amphibians and reptiles on land, such as the crocodile-like phytosaurs.

We are really lucky in the United Kingdom: we have some fantastic surface rocks from many key parts of the geological timescale. The end Triassic is no exception. Figure 7.16 shows a section through the end Triassic along the south UK coast just west of Lyme Regis. If you walk the 10 or so kilometres eastwards along the coast from Seaton Down to Lyme Regis, you are passing through millions of years of Earth's history – starting with the lighter rocks of the end-Triassic period, the mass

FIGURE 7.16 Photograph looking north-east of the Triassic–Jurassic transition exposed in the rocks of the cliffs between Pinhay Bay and Lyme Regis, United Kingdom. Photo: Stephen Hesselbo.

extinction, and then into the darker layered Jurassic rocks above. If you were able to analyse these rocks in detail, you would see that this was once a lovely diverse ecosystem, its debris recorded in the layered sediments of the pre-extinction level. Passing through the extinction event, you would see the fossils become much sparser in the sedimentary layers. Finally, on top of these, you would find fossils from a new ecosystem. The beautiful ammonite pavements near Lyme Regis are part of the recovery phase (Figure 7.17).

Rocks from the Triassic–Jurassic transition are exposed at the surface in several other places around the UK, including a fantastic sequence at St Audrie's Bay in Somerset, not far from where I grew up. It is also possible to find rocks covering the end Permian, the period recording the largest extinction event in the geological record – along the coast near Budleigh Salterton in east Devon, for example.[7]

In these coastal sections, you can walk or climb through the rock strata and sample them, with the right equipment and permissions. We get other important samples by drilling down through layers of rock to obtain rock cores, both on land and on the seabed. By looking closely at the fossil remains within the different layers, scientists can understand something of what was happening in the biological realm millions of years ago. The carbon isotope signature (the balance between carbon atoms with slightly different masses) provides information about the carbon cycle. Other indicators reveal trends in global temperatures and rainfall. All this can be recorded in the same rocks, and so tied together in time.

In contrast, the information about the timing of large igneous province volcanism (used in Figure 7.5, Figure 7.8, and elsewhere) does not come from these sediments. Those conclusions were obtained by heading to the lava flows left by the volcanic outpourings and taking rock samples back to the lab. There, we can measure the products of radioactive decay trapped inside the rock's crystals and calculate how old they are. There are measurement uncertainties, however, and different parts of the same rock sample will yield a distribution of ages. Add to this the real differences in eruption dates of different flows across the area and the uncertainty in dating the

---

[7] It does turn out that the Permian–Triassic transition is in the nudist section of the beach. Exposure of a rather less geological variety!

FIGURE 7.17 Ammonite pavement at Lyme Regis, United Kingdom.
Photo: Tamsin A. Mather.

sediment cores, and the extent of the challenge of linking volcanic activity with environmental response becomes apparent. In rare cases, such as the end-Triassic rocks in parts of North America and Morocco, we are in the fortunate position of having some of the Central Atlantic Magmatic Province lava flows interleaved with sedimentary columns recording the extinction. This certainly makes it easier to connect the igneous and stratigraphic timescales, but it also shows that there were active flows before and after the mass extinction. Hence, to understand the link between volcanism and biology, we need to clarify when the largest pulses of lava and gas were emitted on the province scale; it does not suffice to consider individual flows that happen to be captured in particular cores or outcrops.

New Insights into Old Enigmas?

One way of making progress on this problem is to turn to lessons learnt from studying active volcanoes, and to look for the sort of fingerprints that volcanic emissions might have left behind in the sedimentary rocks, sometimes deposited at considerable distance from the volcanic flows. This is where measurements that we started making for completely different reasons show promise. I have been interested in volcanic emissions of the element mercury since the first year of my PhD. Mercury is toxic, and human activities like burning fossil fuels put it into the environment. Understanding its natural fluxes is important to get a picture of the planet's natural mercury cycle.

Mercury is a really unusual metal. I used to secretly enjoy it when someone broke one of the mercury thermometers in a school science lesson – the glistening balls of quicksilver scattering across the worktop. The use of mercury thermometers has been phased out due to the element's toxicity, and since 2013 human release of mercury has been regulated by an international treaty known as the Minamata Convention. Mercury is the only metal that is liquid at room temperature, which means it also exists more readily as a vapour than other metals. Mercury vapour is in the background air all around us. If you take a deep breath, your lungs are exposed to 10 or so picograms of mercury. The work I did with colleagues around the time of my PhD – making measurements of mercury in volcanic emissions like those from Masaya and Etna – confirmed that volcanoes are a major natural source of mercury. Unlike most other metals, it is largely emitted as a gas and thus widely dispersed in the atmosphere.

Starting with the measurements taken by a Canadian research group of sediments in proximity to the Siberian Traps, scientists have started to notice spikes in the levels of mercury in periods coinciding with dated large igneous provinces. This is an exciting development. If we can show that these mercury spikes are indeed the fingerprints from periods of intense volcanism on the province scale, it means that we no longer have to rely on knitting together the sedimentary and igneous timescales, with all their complexities and uncertainties. Instead, the volcanism will be recorded in the same sediment samples that record the changes in Earth's environment and biology.

Figure 7.18 shows an example of our own work on the end Triassic. You can see how the mercury concentration normalised with respect to total organic carbon (Hg/TOC)[8] increases at the end-Triassic extinction, exactly when the carbon isotopes tell of a perturbation in the carbon cycle. This suggests a burst of intense Central Atlantic Magmatic Province volcanism coincident with the biological and environmental changes recorded in the sediments.

One of the things that particularly excites me about this period, where we have lava flows interleaving with the sediments themselves, is that we can even speculate about which flow units this mercury spike originated from. At about the time of the end-Triassic extinction, we know from the sediment cores in Morocco that the Lower-High Atlas formation was being emplaced. It is awe inspiring that over 200 million years ago, these massive lava flows in present-day Morocco potentially bled out mercury from the Earth's interior that got transported through the atmosphere, hydrosphere, and biosphere and became locked up in sediments that were eventually sampled by scientists in Austria. The samples reached my lab in Oxford and were heated to 700 °C, so that the atoms of mercury were released and analysed, yielding a point on a graph telling us something about the world of 200 million years ago.

[8] This normalisation is necessary because of the role organic carbon plays in fixing mercury.

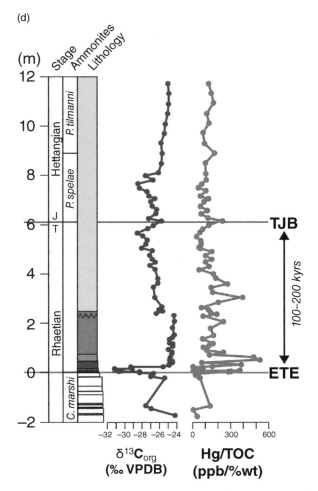

FIGURE 7.18 Mercury data from Kuhjoch, Austria, with accompanying biostratigraphy, lithology, and carbon isotope data. Source: L. M. E. Percival, M. Ruhl, S. P. Hesselbo, H. C. Jenkyns, T. A. Mather et al., 'Mercury evidence for pulsed volcanism during the end-Triassic mass extinction', *Proceedings of the National Academy of Sciences*, 114 (2017), 7929–7934 (at 7931, Fig. 2A), doi:10.1073/pnas.1705378114.

There is much to understand about the processes that control these mercury signals in the geological record, but this new tool could tell us a lot about the causes and effects at play many hundreds of millions of years ago. It also means that we might be less constrained by the preservation of the large igneous province rocks themselves. Others and I have now started to explore how these techniques can unlock new information about events deeper back in Earth's history, like the end-Ordovician mass extinction.

Final Remarks

I have come to the end of the journey I wanted to take you on. To finish, let us return to Figure 7.1, with me staring into the mouth of a volcano. I hope to have convinced you that, by peering into volcanoes and thinking about the enigmas they present, we get insights not only into the mystery of what our planet is doing beneath our feet, but also into Earth's deep history, including the history of life itself. Volcanoes have shaped our world. You would not be sitting where you are today, reading this volume and pondering the many enigmas explored within, without our planet's generation upon generation of volcanoes.

8 The Enigma of Mind: A Theory of Evolution and Conscious Experience

ALBERT Y.-M. LIN AND J. DEREK LOMAS[1]

Introduction

In July 2011, while surveying the buried remains of a cluster of fourteenth-century ceremonial structures in the remote mountains of Northern Mongolia – in the heart of a region known as the Ich Korig (the 'Forbidden Zone'), and at the base of a mountain believed to be the historic Burkhan Khaldun, or 'God Mountain' – my team and I encountered a group of seven Mongolian shamans who had come on pilgrimage to commune with that very location. We had travelled for three days by horseback and heavy four-wheel-drive trucks through the rugged wilderness. Amidst rains and mudslides, we forded swollen rivers to get to the mountain. Foreigners have been restricted from this specific location for over 800 years by decree of Genghis Khan himself, and even though we were operating with special permissions from the Ministry of Culture, that night I was summoned to meet the head shaman in a moment of reckoning.

> Field notes July 18th, 2011, 12:50 am – I will struggle to find the words to describe what I just experienced and although I will surely fail to do it justice, I must try. My heart beat loudly in my throat as the hammer fell upon the drum, its rhythm pounded through the air, like waves through the thick fog. As the sunlight hit the amber tone the mist began to rise from the knoll above our camp. The shaman had summoned me, Ishdorj had guided me to sit next to him, and to recite the respected words that sounded like 'amaraa sanbano uhau'. I was terrified I would not say them

[1] Note to the reader: as this chapter deals directly with conscious experience, we take the unconventional approach of weaving between first-person accounts and a collaborative scholarly discussion. The first-person accounts are told from Albert Lin's perspective.

correctly. The beat moved faster, then faster still until it almost seemed like a constant burst of energy.

Then the shaman, cloaked with a black facial mask of thick cloth covering his eyes, rose to his feet and spun in circles, slamming his hammer into the drum in an explosion of sound, his long cords of cloth (blue, black, white, maroon) spun in the air around him. And then, suddenly, silence. He dropped to the ground in a sitting position and leaned forward while jerking his head back and forth like a crow. His voice grunted low, in spurts of three. As if he were clearing his throat, but also as if he were agreeing with some internal message. 'Unch, unch, unch.'

Throughout my life I have had the great privilege of travelling the world and, upon reflection, I realise that I have encountered aspects of shamanism in many distant cultures. For instance, deep within the Pacific Ocean on the island of Pohnpei – home to the capital of Micronesia – we obtained permission to enter and apply aerial LIDAR to survey the enigmatic archaeological site of Nan Madol through our participation in a tribal ceremony with the Nahnmwarki (High Chief) of the Madol En Inal clan. Like most ceremonies in this oceanic culture, it began with the sacred and mind-altering sakau[2] drink 'to commune with the spirits'. This tradition of ritual altered states is so embedded into the cultural story of the Pacific that not only are sakau pounding stones found within the heart of the ancient megalithic city of Nan Madol (an artificial island complex built *c.* 1180 AD[3]), but the tradition is also within the modern flag of Pohnpei – which carries the sakau cup at its centre. One can easily imagine the importance of rituals of the mind in building an oceanic society brave enough to stand at the shore and harness the collective courage to set sail towards an unknown horizon. To this day, the same bell-like sound – of rock pounding against the sacred sakau stone – rings through mangroves on islands across the Pacific almost daily, summoning all within earshot to begin the sacred ceremony.

[2] Kava (*piper methysticum*).
[3] M. D. McCoy, H. A. Alderson, R. Hemi, H. Cheng, and R. L. Edwards, 'Earliest direct evidence of monument building at the archaeological site of Nan Madol (Pohnpei, Micronesia) identified using [230]Th/U coral dating and geochemical sourcing of megalithic architectural stone', *Quaternary Research*, 86(3) (2016), 295–302.

Looking further back into our shared human story, while travelling through Norway's most northern tip, I was stunned by the magnificent Neolithic rock art at Alta. In a remarkable portrait of the emergence of modern culture, thousands of images from as early as 4200 BC are carved into the coastal bedrock of the Alta fjord. In the midnight sun, when long shadows are cast over the subtle contours, I saw ancient representations of human ingenuity, such as deep-sea fishing, and representations of our collective spiritual roots. The site's UNESCO World Heritage designation describes the 'exceptionally high number of human figures and compelling portrayals of prehistoric social life, dancing, processions, and rituals'. They seem to show 'communication between the world of the living and the worlds of the spirits [giving] insight into the cosmology of prehistoric hunters and gatherers'.[4]

All human cultures have developed technologies to shape and explore our conscious experience: ritual, ceremony, art, music, and concoctions of intoxication. These technologies of consciousness have been woven into our societies, leaving temples and monuments in their wake. Why? How have these consciousness-altering tools affected human evolution?

From Anthropology to Personal Experience

In the most unexpected way, my interest in the anthropological enigma of consciousness intersected with my own personal evolution. Five years after standing on Mongolia's most sacred mountain, I found myself lying in the dirt: blood pouring out of my leg, I had been crushed by the weight of an overturned vehicle. Rushed to the hospital, I unwittingly embarked upon a journey into the enigma of the mind.

The following month I endured the intense mental and physical pain of an attempt at limb salvage. The repeated surgeries each brought rounds of excruciating wound care, hours in a hyperbaric oxygen chamber, and an intimacy with the overall experience of pain that I had not known possible. Ultimately it didn't work, and a final surgery removed the material aspects of a limb that had been a part of my mental body map for over 36 years. But the imprint of the pain in my mind endured.

[4] whc.unesco.org/en/list/352/.

I cannot recall if it was just moments or days after the amputation, but gradually I began to notice a tingly sensation in my right foot – the one that was no longer there. The phenomenon of phantom pain soon followed. Burning, darting, pangs of pain, the sensation of every tendon, ligament, and bone in my ankle and foot being repeatedly broken and lit on fire in waves of electric shock coursing through the air where my foot once was. This pain was real, yet it occurred in a part of my body that I rationally knew did not exist.

Serendipity's hand revealed that I happened to work at the same university campus as Dr V. S. Ramachandran, the man who literally wrote the book on the phantom limb phenomenon.[5] In our first meeting, he asked: "Do we really know where the body ends and the mind begins?" A long-time distinguished professor at UC San Diego, 'Rama' had developed the use of 'mirror box therapy'. Placing a borderless mirror between my legs, he could 'trick' my mind into seeing an illusion of a functioning right leg in the reflection of the left. This experience permits new (and less painful) memories to be created; memories to displace my neural attachments to pain and trauma. The mirror box therapy was uncanny: I could see my right toes wiggling again, a mirror image always moving in synchrony with my left. It was comical but comforting; somehow the visual experience in my mind gave me physical relief. Yet, almost as soon as the mirror was removed, the pain would return. The stubbornness of my mind (the 'ego' through which I perceive the world) would not hold on to the new narrative I was attempting to tell my brain.

Now incapacitated by the weight of the mysterious 'phantom pain', I needed to find a solution. Suspecting that a lack of neuroplasticity could be the culprit, I began exploring cognitive technologies that I had encountered throughout the world: Kundalini Yoga, chanting, meditation, sensory deprivation float tanks, among others. Finally, deep in the sand dunes of Death Valley, I discovered that a single large dose of psilocybin or 'magic mushrooms' – in conjunction with Ramachandran's mirror therapy – finally resolved my phantom limb pain in an enduring way. Somehow, the ego-crushing power of the hallucinogenic experience enabled my brain to

---

[5] V. S. Ramachandran and S. Blakeslee, *Phantoms in the Brain: Probing the Mysteries of the Human Mind* (New York, NY: William Morrow & Co., 1998).

loosen up and accept the re-framed reality that the mirrors presented. This case has now been published and a formal clinical trial of psilocybin in phantom pain therapy is under way at UC San Diego, at the newly formed Psychedelics and Health Research Initiative.[6]

In phantom limb pain, the mind is clearly capable of crafting realities that extend beyond what we can rationally understand. No amount of verbal explanation could convince my brain to let go of the pain it felt was real. But then a few grams of entheogenic mushrooms facilitated an experience so overwhelming that I lost my sense of self. In that moment of communion with a vast cosmos, I was able to let go of the bounds of prior truths and re-imagine a more harmonised state of existence. It was similar to what Mihaly Csikszentmihalyi described as a 'flow state'.[7] Had I been more skilled, perhaps I could have achieved a comparable experience through meditation or another practice. So much evidence now shows that psychedelics can increase neural plasticity that some scientists have named these drugs 'psychoplastogens'. But psilocybin doesn't just seem to create a biochemical effect – it isn't aspirin. The *experience* itself seems necessary for the benefit of plasticity.[8]

Ramachandran once suggested that our species could more appropriately be named *Homo plasticus* because 'lifelong plasticity' is the 'central player in the evolution of human uniqueness'.[9] In my case, *psilocybin* augmented my plasticity, but I suspect that many non-drug-induced states of consciousness could also produce neural plasticity.[10] Perhaps

[6] V. S. Ramachandran, C. Chunharas, Z. Marcus, T. Furnish, and A. Lin, 'Relief from intractable phantom pain by combining psilocybin and mirror visual-feedback (MVF)', *Neurocase*, 24(2) (2018), 105–110.

[7] 'Flow is the way people describe their state of mind when consciousness is harmoniously ordered', M. Csikszentmihalyi and J. Nakamura, 'Flow, altered states of consciousness, and human evolution', *Journal of Consciousness Studies*, 25 (11–12) (2018), 102–114.

[8] C. Ly, A. C. Greb, L. P. Cameron, J. M. Wong, E. V. Barragan et al., 'Psychedelics promote structural and functional neural plasticity', *Cell Reports*, 23(11) (2018), 3170–3182; N. R. P. W. Hutten, N. L. Mason, P. C. Dolder, E. L. Theunissen, F. Holze et al., 'Low doses of LSD acutely increase BDNF blood plasma levels in healthy volunteers', *ACS Pharmacology & Translational Science*, 4(2) (2021), 461–466.

[9] V. S. Ramachandran, *The Tell-Tale Brain: A Neuroscientist's Quest for What Makes Us Human* (New York, NY: W. W. Norton & Co., 2011).

[10] In 2019, Robert Turner, director at the Max Planck Institute for Human Cognitive and Brain Sciences in Leipzig, Germany, published a new theory

even the shamanic experience I encountered in the 'Forbidden Zone' was accompanied by a similar kind of neural transformation?

Human Experience, Evolution, and Design

I now wish to introduce my co-author, Dr James Derek Lomas, a professor at Delft University of Technology. As a cognitive scientist, designer, and philosopher, he explores human experience from within the field of human-centred design. Over the last decade, Dr Lomas has been my closest friend and intellectual partner. The journey I described above was largely accompanied by our conversations surrounding consciousness and its many augmentations. Throughout our adventures together we have explored the intersection of human experience, technology, and philosophy. The present chapter shares the culmination of our discussions about the role of conscious experience in evolution – and for this, we will now transition from a first-person narrative into a joint discussion.

Darwinism in the Mind

'It is a great honor to have been invited to give the first Darwin Lecture at Darwin College, in Cambridge, which of all universities is most closely connected with Charles Darwin and the Darwin family.'

So began the philosopher Karl Popper in the inaugural Darwin Lecture, 'Natural Selection and the Emergence of Mind'.[11] In his accompanying book chapter, Popper proposed a new idea: that the evolutionary forces of natural selection were actually operating within the conscious mind. He called this 'the greatest marvel of our universe'. As an example,

---

regarding the role of neural plasticity in ritual experiences. He suggested that dramatic rituals could help catalyse equally dramatic consolidation in neural pathways, particularly when involving rhythmic entrainment. See R. Turner, 'Finding likeness: Neural plasticity and ritual experience', *Anthropology Today*, 35 (3) (2019), 3–6.

[11] Karl Popper is well known for his advocacy of falsifiability in the sciences – that hypotheses must be able to be proven false – but he also said that 'the theory of natural selection could be untestable (as is a tautology) and yet of great scientific interest'. Popper additionally said, 'the doctrine of natural selection is a most successful metaphysical research program'. See K. Popper, 'Natural selection and the emergence of mind', *Dialectica*, 32 (1978), 339–355.

Popper described how Albert Einstein generated an 'immense number of hypotheses' and then rejected the vast majority, in a process of evolutionary natural selection. As another example of evolution through natural selection, Popper described his own experience of writing: both writing many words and crossing many out. He then evoked the image of an artist, painting out spots of colour before reflecting whether to keep the strokes or paint over them again. For Popper, this was the everyday nature of evolutionary forces manifesting in the mind – the experience of evolution. He proposed that evolutionary forces occur within three 'mutually interacting' worlds: the physical world, the world of experience, and the world of artefacts.

We now wish to further consider Popper's notion of evolutionary forces in human experience and extend the idea to more extreme and ancient human circumstances. We are inspired by Popper's call for 'intellectual daring in the search for truth' while remembering his caution to avoid making claims about ultimate questions, riddles of existence, or humanity's task in this world. We humbly seek to understand how conscious experience might relate to evolutionary theory and are motivated by the very same question that motivates all thinking on the topic of evolution: how did we arrive at the world we see today?

Summary: A Theory of Evolutionary Forces in the Mind

The elegant sculpting power of natural selection is not limited to genetic systems.[12] Instead, evolutionary fitness forces are expected to emerge within any system that has three key properties: variation, replication, and selection. To understand the operation of evolution in the mind, we will review three properties of experience that seem to satisfy these three requirements of evolutionary theory. There may be many properties of experience that can support variation, replication, and selection, but we focus on the following: (1) variation via imagination, (2) replication via sympathy, and (3) selection via harmonisation.

---

[12] R. Bondurianskyy and T. Day, 'Nongenetic inheritance and its evolutionary implications', *Annual Review of Ecology and Systematics*, 40 (2009), 103–125.

We begin with the topic of human imagination as an important source for diversity and variety in the human mind. After reviewing key archeological milestones in the development of the creative imagination, we reflect on the curious cultural ubiquity of shamanistic practices and consider the basic drive for altering consciousness. We suggest that altered states of consciousness may facilitate neural plasticity and produce increased variation in the imagination. If so, then the uneven geographical distribution of psychoactive plants may have influenced the diversity of human culture, both in the recent past and deep into prehistory.[13]

After imagination, we suggest that the human capacity for sympathy plays an important role in the replication of conscious experiences between people. Specifically, we consider the theory that *sympathetic resonance* serves as a prelinguistic mechanism for the replication of conscious experience from one person to the next.[14] The capacity to replicate feelings through sympathetic resonance may have emerged even prior to deliberate human imitation. Ramachandran's theory regarding the role of 'mirror neurons' in cultural evolution is one description of how neural resonance might create the capacity to share mental states and experiences across people and time. With evidence from recent multi-person brain-imaging 'hyperscanning' studies, we link the profoundly rhythmic nature of the human brain to the reproduction of experiences between people – through resonance. Finally, we give examples of cultural forms that may have evolved to maximise human resonance and sympathy, such as Mayan bloodletting ceremonies.

Having introduced sources for the variation and replication of experience, we then consider processes that might govern Darwinian selection forces in the mind, namely within the competition for attention. In what might be described as an integration of Platonism with evolutionary theory, we focus on *harmonisation* as a selection pressure and fitness function in the mind.[15] Both Darwin and Alfred Russel Wallace gave descriptions of this 'harmony' in natural selection. We show how this classical concept plays a role in the

---

[13] For current distribution maps of peyote cactus (*Lophophora williamsii*), fly agaric mushroom (*Amanita muscaria*) and mushrooms of the genus Psilocybe, see © Open Street Map and GBIF contributors www.gbif.org/.

[14] J. D. Lomas, A. Lin, S. Dikker, D. Forster, M. L. Lupetti et al., 'Resonance as a design strategy for AI and social robots', *Frontiers in Neurorobotics*, 16 (2022), article 850489.

[15] J. D. Lomas and H. Xue, 'Harmony in design: A synthesis of literature from classical philosophy, the sciences, economics, and design', *She Ji: The Journal of Design, Economics, and Innovation*, 8(1) (2022), 5–64.

mind, both as a metaphor and as a mechanism. We then summarise the powerful role that harmony has played in empirical science, past and present.

In the end, we hope to have brought new detail to Popper's fascinating link between evolution and conscious experience. Are these ideas practical? We conclude by considering how new technologies and cultural practices might build upon these principles to support a more positive and resilient human future. To illustrate this very possibility, Dr Albert Lin shares a final first-person account of his journey to a secret mystical temple within the heart of the jungles of the Yucatán peninsula.

The Mind and Evolution

Despite living thousands of miles apart, Charles Darwin (1809–1882) and Alfred Russel Wallace (1823–1913) each independently discovered the principles of evolution. Darwin wrote 'I never saw a more striking coincidence.'[16] Both Darwin and Wallace also considered the role of the human mind in evolution, with Wallace suggesting that 'it may well be that evolution is a fundamental law of the universe of mind as well as that of matter'.[17] It is this enigma that we investigate: the nature of evolutionary forces operating *within* the realm of the conscious mind – and how evolution in the mind manifests itself in the remarkable diversity of human cultures.

Imagination, Variation, and Human Evolution

Darwin and Wallace agreed that once humans developed certain mental abilities, evolution would no longer affect the body so strongly, but that evolution would continue *within the mind*. Indeed, while *Homo sapiens* emerged between 350,000 and 260,000 years ago,[18] the earliest known representational cave painting, found in Indonesia, dates to 43,000 BC.[19]

[16] C. Darwin, A. R. Wallace, G. Sarton, C. Lyell, and J. D. Hooker, 'Discovery of the theory of natural selection', *Isis*, 14(1) (1930), 133–154.
[17] A. R. Wallace, 'The harmony of spiritualism and science', *Light* (25 July 1885).
[18] C. M. Schlebusch, H. Malmström, T. Günther, P. Sjödin, A. Coutinho et al., 'Southern African ancient genomes estimate modern human divergence to 350,000 to 260,000 years ago', *Science*, 358(6363) (2017), 652–655.
[19] This painting was discovered not far from where Wallace first experienced the incredible insight of evolution through natural selection. See A. Brumm, A. A. Oktaviana, B. Burhan, B. Hakim, R. Lebe et al., 'Oldest cave art found in Sulawesi', *Science Advances*, 7(3) (2021), article eabd4648.

The first known European representational art dates to 35,000 BC and the earliest discovered in Africa dates to about 30,000 BC.[20] Thus, while the human body evolved in Africa and dispersed across the world, it seems that evolution in the mind took place in locations all around the world. What evolved, in each case, seems to be a profound awakening in the human capacity for imagination.

The archeological record can indicate key threshold events in the evolution of the imagination.[21] For instance, the Löwenmensch or Lion-Man sculpture (c. 35,000 BC) (as depicted earlier in this volume in Figure 1.1) clearly shows evidence of mentally synthesising human and animal parts in the imagination and carving them into reality. The emergence of 'synthetic imagination' seems to have given humans the capacity to mentally model different objects and consider their combination through trial and error in the mind alone. This novel capacity for synthetic imagination relies on the ability to literally synchronise different object-encoding neural circuits together – a process known as prefrontal synthesis.

Once humans could play around with combinations of objects in their mind, rather than through physical manipulation, cultural evolution may have accelerated dramatically. Complex artefacts like boats (enabling the peopling of Australasia), traps (for killing the soon-to-be-extinct megafauna), or other game-changing tools and techniques could now be imagined. These factors alone would create an enormous evolutionary advantage. But the emergence of the Synthetic Imagination also seemed to enable activities with much less obvious benefit for Palaeolithic humans, like rituals, ceremonial burial, music, and art.

Near the discovery of the Löwenmensch, researchers uncovered the highly sexualised Venus of Hohle Fels (Figure 8.1) and the earliest known bone flutes (Figure 8.2), found just 70 centimetres apart.[22] This

[20] R. Ego, *Visionary Animal: Rock Art from Southern Africa*, trans. D. Dusinberre (Johannesburg: Wits University Press, 2019).
[21] A. Vyshedskiy, 'Neuroscience of imagination and implications for human evolution', *Current Neurobiology*, 10(2) (2019), 89–109.
[22] N. J. Conard, 'A female figurine from the basal Aurignacian of Hohle Fels Cave in southwestern Germany', *Nature*, 459(7244) (2009), 248–252; N. J. Conard, M. Malina, and S. C. Münzel, 'New flutes document the earliest musical tradition in southwestern Germany', *Nature*, 460(7256) (2009), 737–740.

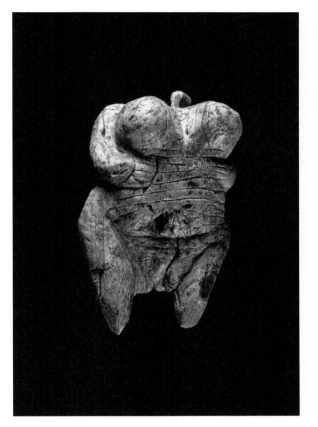

FIGURE 8.1 The Venus of Hohle Fels. Photo: Hilde Jensen. Copyright: Universität Tübingen.

set of artefacts shows the emergence of a capability to shape the material environment for the purpose of *evoking experiences* in other people.[23] These magnificent artifacts do not serve a practical purpose as a spear or a flint would, but instead describe an awakening of the artist in our human nature, an aspect of the mind that is driven by our uniquely existential consciousness.

---

[23] The much older use of ochre may also be evidence of this. See T. Hodgskiss, 'Ochre use in the middle stone age', in *Oxford Research Encyclopedia of Anthropology* (Oxford: Oxford University Press, 2020), online.

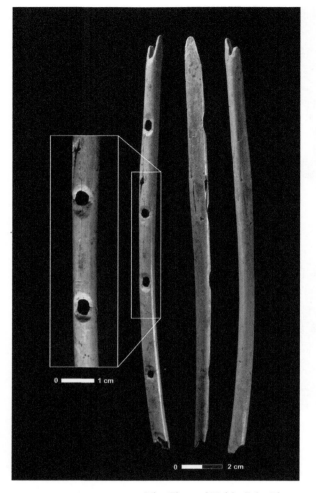

FIGURE 8.2 The Flute of Hohle Fels. Photo: Hilde Jensen. Copyright: Universität Tübingen.

Does the birth of the artist in humanity coincide with the evolution of spirituality in human culture? Or is the art simply the evidence of a far more ancient set of spiritual practices? This brings us to shamanism, an ancient and mysterious phenomenon associated with the performative induction of spiritual experiences in the mind.

Shamanism, Altered States and Diversity in the Human Imagination

The term 'shaman' comes from the Siberian Tungus people.[24] But it is used to describe phenomena that are present in practically all cultures around the world, ancient and modern,[25] where altered states of consciousness are used to conduct divinatory work, prayer, healing, or other soul journeys, manipulations, or transformations. Shamanism is notoriously difficult to define with precision, but it involves the widespread tendency for specialised individuals to engage in what distinguished archaeologist Johan Reinhard described as a 'non-ordinary psychic state'.[26]

In animal evolution, the presence of a common phenotype or characteristic suggests either a common ancestor or convergent evolution. Similarities between disconnected shamanic traditions may therefore point to a common lineage or to convergent evolution due to common biological factors. For example, accelerating drumming, reaching 4 Hz or 240 beats per minute, is prevalent in shamanistic tradition globally and may have evolved due to interactions with natural theta frequencies of brain response.[27] Global similarities in shamanistic practice may be a result of ancient origins as well as the convergence of cognitive technologies after generations of trial and error, refinement and integration.

Apart from shamanism, pathways to altered states of consciousness are found throughout cultures globally. Across a sample of 488 world societies, over 90 per cent exhibited some form of institutionalised altered states of consciousness.[28] In Islamic Sufism, for example, Whirling Dervishes

---

[24] S. M. Shirokogoroff, *Psychomental Complex of the Tungus* (London: Routledge and Kegan Paul, 1935).

[25] M. Eliade, *Shamanism: Archaic Techniques of Ecstasy* (New York, NY: Pantheon Books, 1964); D. Stern, 'Masters of ecstasy', *National Geographic*, 222(6) (2012), 110–131.

[26] J. Reinhard, 'Shamanism and spirit possession – the definition problem', in J. T. Hitchcock and R. L. Jones (eds.), *Spirit Possession in the Nepal Himalayas* (Warminster: Aris and Phillips, 1976), pp. 12–20; M. Singh, 'The cultural evolution of shamanism', *Behavioral and Brain Sciences*, 41 (2018), 1–83.

[27] M. J. Hove, J. Stelzer, T. Nierhaus, S. D. Thiel, C. Gundlach et al., 'Brain network reconfiguration and perceptual decoupling during an absorptive state of consciousness', *Cerebral Cortex*, 26(7) (2016), 3116–3124.

[28] E. Bourguignon, 'World distribution and patterns of possession states', in R. Prince (ed.), *Trance and Possession States* (Montreal: R. M. Bucke Memorial Society, 1968), pp. 3–34.

attempt to dissolve one's ego by entering a trance state via the ritual of spinning continuously amidst chants and song. Hindu Sadhus devote themselves to meditation, a life of ritual and a renunciation of worldly things to achieve a certain state of mental experience. Gospel choirs entrance churchgoers to prepare them for prayer, preaching, and liturgy.

Cultures have developed many techniques for inducing trance-like altered states, including drumming, chanting, dancing, meditation, sexual practices, sacred objects, architecture, immersion in nature, pain, fasting, music, extreme sports, sleep deprivation, and the use of psychoactive substances, to name but a few. Yet, with the exception of sexual trance (which is rarely institutionalised),[29] few of these methods seem to offer much evolutionary benefit to the individual participants. The pursuit of altered states might even seem maladaptive, particularly for Neolithic peoples living in a dangerous environment. Yet, the tendency exists nearly universally across cultures. What benefits do altered states bring to people, now or deep in the past?

In an fMRI study, researchers observed that shamanic trance states were associated with a suppression of perceptual stimuli. The authors propose that shamanic trance enables a decoupling of cognition from the constraints of the external world. Through dissociation, fundamental aspects of reality might be grasped in different ways. This decoupling may be important for exploring the diverse and ineffable experiences deep in one's own imagination. Visionary shamanic techniques may have offered the benefit of enabling practitioners to see new realities in their imagination that others could not.[30]

Altered States as a Source of Variation in the Mind

For natural selection to operate in the realm of the mind, it needs a source of variation. Variation in conscious experience can be produced in many ways, including creative play and dreams. We propose that some altered states of consciousness are a culturally important mechanism for

[29] A. Safron, 'What is orgasm? A model of sexual trance and climax via rhythmic entrainment', *Socioaffective Neuroscience & Psychology*, 6(1) (2016), 1–17.

[30] R. Noll, 'Mental imagery cultivation as a cultural phenomenon: The role of visions in shamanism', *Current Anthropology*, 26(4) (1985), 443–461.

promoting the generation of diversity and variation in mental experiences. For evidence, we look to the clinical research and neuroimaging emerging from the current renaissance in psychedelic research in medicine. In laboratory studies, psychedelics have been shown to promote divergent thinking[31] and associative processing[32] – and the vividness and richness of the imagination is a key self-reported characteristic of psychedelic experiences. Researchers describe the subjective effects of psilocybin or magic mushrooms as 'broadly unconstrained perception and cognition, hyper-associative cognition and, at higher doses, a breakdown in the perception of time, space and selfhood'.[33]

Heavy drumbeats, ritual, or psychotropic experiences can disrupt the 'default mode' of our logical and verbal mind.[34] Once a person's dominant brain narrative releases control, there emerges an opportunity for bottom-up elements to constitute a more diverse range of thinking. This is the realm of intuition and free play in the imagination. Disrupting inhibitory control can create an experience that more broadly explores the space of cognitive combinations, leading to the discovery of new ways of seeing the world.

This appears to be reflected on a neurological level. A recent placebo-controlled fMRI study showed the significant causal impact of psilocybin on connections between previously unconnected regions of the brain (Figure 8.3).[35] The authors note 'the emergence of strong, topologically long-range functional connections that are not present in a normal state'. They conclude that the brain imaging shows 'a less constrained and more

[31] N. L. Mason, E. Mischler, M. V. Uthaug, and K. P. C. Kuypers, 'Sub-acute effects of psilocybin on empathy, creative thinking, and subjective well-being', *Journal of Psychoactive Drugs*, 51(2) (2019), 123–134.

[32] M. Spitzer, M. Thimm, L. Hermle, P. Holzmann, K. A. Kovar et al., 'Increased activation of indirect semantic associations under psilocybin', *Biological Psychiatry*, 39(12) (1996), 1055–1057.

[33] L.-D. Lord, P. Expert, S. Atasoy, L. Roseman, K. Rapuano et al., 'Dynamical exploration of the repertoire of brain networks at rest is modulated by psilocybin', *Neuroimage*, 199 (2019), 127–142.

[34] I. McGilchrist, *The Master and His Emissary: The Divided Brain and the Making of the Western World* (New Haven, CT: Yale University Press, 2019).

[35] G. Petri, P. Expert, F. Turkheimer, R. Carhart-Harris, D. Nutt et al., 'Homological scaffolds of brain functional networks', *Journal of the Royal Society Interface*, 11(101) (2014), article 20140873.

(a) (b)

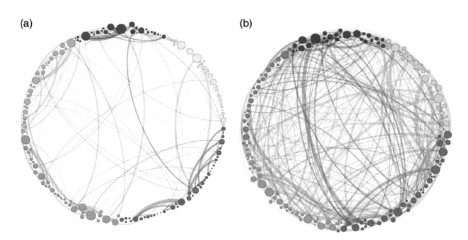

FIGURE 8.3 Simplified visualisation of homological scaffolds representing a comparison of communication between brain networks in people given a non-psychedelic placebo compound (a) or psilocybin (b). Source: G. Petri et al., 'Homological scaffolds of brain functional networks', *Journal of the Royal Society Interface*, 11 (2014), article 20140873, Fig. 6, doi:10.1098/rsif.2014.0873. Licence: CC-BY.

intercommunicative mode of brain function, which is consistent with descriptions of the nature of consciousness in the psychedelic state'.

One recent theory regarding psychedelic action in the brain focuses on the relaxation of existing belief structures. A remarkable fMRI study by Carhart-Harris et al. showed a deactivation of what is referred to as the 'default mode network' (DMN) – a group of brain regions associated with introspection and self-referencing – during the hallucinogenic state produced by psilocybin.[36] The DMN is a normal daily part of brain activity; as we mature with experience, the pathways of information exchange in the brain become more rigid or constrained in their patterns. These patterns in the DMN shape the way in which we perceive the world, quite literally forming the default mode of our self-narrative, or 'ego'. The authors suggest 'the subjective effects of psychedelic drugs are caused by

[36] R. L. Carhart-Harris, D. Erritzoe, T. Williams, J. M. Stone, L. J. Reed et al., 'Neural correlates of the psychedelic state as determined by fMRI studies with psilocybin', *Proceedings of the National Academy of Sciences*, 109(6) (2012), 2138–2143.

decreased activity and connectivity in the brain's key connector hubs, enabling a state of unconstrained cognition'.

In part by reducing inhibitory frontal control, deactivating the DMN, and momentarily turning off our 'ego', a psychedelic experience can allow the imagination to operate more loosely, taking into account a broader variety of bottom-up information streams and producing a greater diversity of mental variation. To describe this effect, Robin Carhart-Harris and Karl Friston[37] introduced the REBUS model, which claims that psychedelics cause an increase in bottom-up information flow and an increase in overall brain signal diversity.[38] The result is that the brain's 'free energy landscape' is made more accessible. Psychedelics are compared to annealing in metallurgy, where heating a system allows it to 'attain a state of heightened plasticity, in which the discovery of new energy minima . . . is accelerated'. The authors further suggest that increased plasticity during psychedelic states enables the emergence of new belief structures that 'resonate more harmoniously with previously hidden or silenced information'.

Ethnobotany and the Geography of Imagination

This next section considers the following question: how might the availability of psychedelic plants and fungi have affected early societies and cultural evolution?

One source of human cultural diversity is geographical diversity. For instance, the local temperature of the environment can influence clothing, shelters, and food. Geographical variations also seem to create an imprint in the imagination: people tend to *experience* different things in a lush rainforest, in a sparse highland steppe, or in a windswept fjord. Another way in which geography may affect the imagination is through variation in the availability of psychoactive plants. Harvard ethnobotanist

---

[37] Friston is famous for devising a now dominant model of predictive coding in the brain known as 'free energy minimisation' which is mathematically equivalent to Smolensky's 'harmony maximisation' function in artificial neural networks. See K. J. Friston and K. E. Stephan, 'Free-energy and the brain', *Synthese*, 159(3) (2007), 417–458.

[38] R. L. Carhart-Harris and K. J. Friston, 'REBUS and the anarchic brain: Toward a unified model of the brain action of psychedelics', *Pharmacological Reviews*, 71(3) (2019), 316–344.

Dr Richard Schultes ('father of ethnobotany') and Dr Albert Hofmann (the first to synthesise LSD and psilocybin) paint a clear picture of the incredibly broad cultural use of these plants throughout the ages:

> The use of hallucinogenic or consciousness-expanding plants has been a part of human experience for many millennia,[39] yet modern Western societies have only recently become aware of the significance that these plants have had in shaping the history of primitive and even of advanced cultures.[40]

The deliberate use of mind-altering plants can be found across the animal kingdom.[41] The cultural use of 'psychoplastogens' probably antedates the emergence of *Homo sapiens* (and may have catalysed it, as has been suggested in the 'Stoned Ape' theory of human evolution).[42] However, ready access to psychoactive plants is highly dependent on ecological conditions. Mind-altering plants such as peyote, San Pedro cactus, ayahuasca, or anadenanthera,[43] for instance, could only have been found in the Americas. By some fluke, the Americas are home to over 200 species of hallucinogenic plants; there are just one-tenth as many psychoactive species across all of Eurasia and Africa. The reason for this imbalance of botanical diversity is an outstanding scientific enigma.[44]

How might the accessibility of psychoactive plants have affected the evolution of the various cultures? The different chemicals found in different plants produce significantly different experiences. Psilocybin mushrooms, *Amanita muscaria*, San Pedro cacti, ayahuasca, datura, kava, or cannabis – they all produce a very different set of mental effects. What might this mean for cultural evolution? If the collective imagination of a

---

[39] Alternatively, this timeline may extend back millions of years, applying the notion that human ancestors may have regularly eaten dung-loving (coprophilic) mushrooms while following the tracks of megafauna, as proposed in T. McKenna, *Food of the Gods: The Search for the Original Tree of Knowledge: A Radical History of Plants, Drugs and Human Evolution* (London: Random House, 1999).

[40] R. E. Schultes, A. Hoffman, and C. Ratsch, *Plants of the Gods: Their Sacred, Healing and Hallucinogenic Powers* (Rochester, VT: Healing Arts Press, 2012).

[41] G. Samorini, *Animals and Psychedelics: The Natural World and the Instinct to Alter Consciousness* (New York, NY: Simon & Schuster, 2002).

[42] McKenna, *Food of the Gods.*

[43] C. M. Torres, 'Archaeological evidence for the antiquity of psychoactive plant use in the Central Andes', *Annali dei Musei Civici Rovereto*, 11 (1995), 291–326.

[44] W. La Barre, 'Old and new world narcotics: A statistical question and an ethnological reply', *Economic Botany*, 24(1) (1970), 73–80.

population is influenced by the types of altered states they can access, then perhaps the local accessibility of certain plants may help explain certain patterns of cultural evolution. For instance, various rock art motifs may have included representations of 'entoptic visuals' – that is, visual patterns from certain altered states may have been carved directly onto the rock.[45] David Lewis-Williams and Thomas Dowson illustrate a comparison between Neolithic art patterns and the neurophysiological visual phenomena sometimes called 'form constants' or 'phosphenes' described by volunteers in a laboratory setting (Figure 8.4). Rock art represents a material record of early human imagination; it is provocative to consider that the history of altered states of mind is reflected in that record (Figure 8.5).[46]

The mere geographical availability of entheogens does not guarantee their cultural adoption. For instance, while several species of psilocybin mushrooms have been widespread in Europe for millennia, scholars agree that there is no definitive textual evidence for the intentional use of psilocybin-containing mushrooms prior to the twentieth century.[47] Only the iconic red and white speckled *Amantia muscaria* mushrooms have any documented ethnographic history (and these mushrooms do not contain the psychedelic chemical psilocybin). Yet, surely 50,000 years of hungry human explorers would have accidentally revealed the power-ful effects of consuming psilocybin mushrooms. Thus, we face an enigma: if psilocybin mushrooms were widely available in Europe, why is there so little remaining evidence of their usage prior to the historical 'discovery' of rituals incorporating magic mushrooms in Mexico in 1957?

[45] J. D. Lewis-Williams and T. A. Dowson, 'On vision and power in the Neolithic: Evidence from the decorated monuments', *Cultural Anthropology*, 34(1) (1993), 55–65.

[46] Alta, Norway, has been designated a UNESCO World Heritage site for its 'compelling portrayals of prehistoric social life, dancing, processions, and rituals'. While many of the (over 6,000) etchings at the site depict human or animal figures, there is also a selection of unexplainable forms, providing support to the theory that some of the art reflects mental or 'entopic' visualisations.

[47] G. Guzmán, J. W. Allen, and J. Gartz, 'A worldwide geographical distribution of the neurotropic fungi, an analysis and discussion', *Annali dei Musei Civici Rovereto*, 14 (1998), 189–280; C. Ruck, 'The mushroom stones: Dionysus, Orpheus and the wolves of war', in D. Spasova (ed.), *Megalithic Culture in Ancient Thrace* (Blagoevgrad: Rilski University Press, 2015), pp. 1–7; G. Samorini, 'Mushroom effigies in world archaeology: From rock art to mushroom-stones', in D. Spasova (ed.), *Proceedings of the Conference 'The Stone Mushrooms of Thrace'* (Alexandroupoli: Greek Open University, 2012), pp. 16–42.

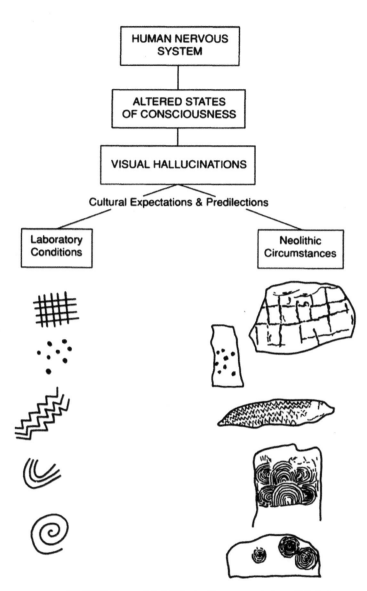

FIGURE 8.4 Model illustrating suggested parallels between neurophysiological visual 'form constants' and Neolithic rock art motifs. Source: J. D. Lewis-Williams and T. A. Dowson, 'On vision and power in the Neolithic: Evidence from the decorated monuments', *Cultural Anthropology*, 34(1) (1993), 55–65 (at 56, Fig. 1, after Siegel 1977 and Shee Twohig 1981). Republished with permission of University of Chicago Press – Journals, permission conveyed through Copyright Clearance Center, Inc.

FIGURE 8.5 Rock art from roughly 4000 BC observed at Alta, Norway.
Photo: Albert Lin.

During our investigations on this unusual matter, we have come across one striking and provocative counter-example which suggests the possibility of representations of the entheogenic use of mushrooms in ancient Europe. With due caution, we note that there are well over 100 prehistoric carvings on Stonehenge that, at least to a naive modern eye, appear 'mushroom-shaped' (Figure 8.6).[48] However, since the carvings were first discovered on Stonehenge in 1953, scholars have agreed that the carvings represent bronze axeheads, each with its blade facing the sky. Even today, this hypothesis remains the most likely. Experts have dated many of the Stonehenge carvings to 1500–1300 BC because this was the only time period in the archaeological record when these unusual 'crescent-shaped' bronze axeheads were used in Britain.

This approach to dating leaves some room for doubt, however. If the development of new technologies for dating rock art were to reveal a

[48] M. Abbott and H. Anderson-Whymark, *Stonehenge Laser Scan: Archaeological Analysis Report* (Portsmouth: English Heritage, 2012). Although described as 'axe-heads', the carvings are far more diverse in form than the comparatively conservative shapes of bronze age axeheads from southern Britain. See S. Needham, *The Classification of Chalcolithic and Early Bronze Age Copper and Bronze Axe-Heads from Southern Britain* (Oxford: Archaeopress, 2018).

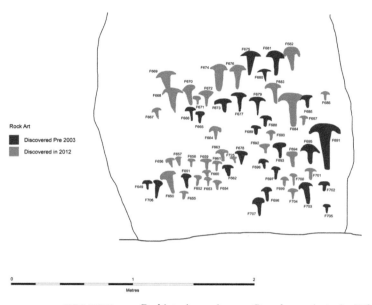

Rock Art

■ Discovered Pre 2003
■ Discovered in 2012

0   1   2

Metres

FIGURE 8.6 Prehistoric carvings on Stonehenge (exterior E face of Stone 4). Source: M. Abbott and H. Anderson-Whymark, *Stonehenge Laser Scan: Archaeological Analysis Report* (Portsmouth: English Heritage, 2012), 29, Fig. 12. © Historic England.

substantially different date for these carvings (e.g., a date in late antiquity), this might weaken the axehead hypothesis. Therefore, an alternative hypothesis may at least be worthy of further investigation: that some carvings on Stonehenge could represent the ancient use of entheogenic mushrooms. Putting aside the debate over what exactly these enigmatic carvings actually represent, we can state with little doubt that the modern 'discovery' of psilocybin mushrooms catalysed a powerful transformation of human culture.[49] Given their geographical availability, it seems highly unlikely that this was the very first time that 'magic mushrooms' had this transformative cultural effect.

Why might psychoactive substances have been used in the distant past and then lost again? Consider *Cannabis sativa*, which is the most commonly used entheogen. Cannabis has a well-documented history in ritual experience, beginning with a rich textual description of Scythian

[49] R. G. Wasson, 'Seeking the magic mushroom', *Life Magazine*, 42(19) (1957), 100–120.

ceremonial cannabis smoke-tents in 430 BC.[50] In 2019, archaeologists found hard evidence for the smoking of cannabis in the mountains of eastern China, dating to 500 BC. There, researchers found braziers with high Δ^9-tetrahydrocannabinol (THC) cannabis in tombs that showed evidence of 'funerary rites that included flames, rhythmic music, and hallucinogen smoke, all intended to guide people into an altered state of mind'.[51] Nevertheless, there are large gaps in the history of cannabis. Despite its having reached Europe by 6000 BC,[52] the psychoactive effects of cannabis seem to have been ignored by the civilisations of Egypt, Greece, Rome, and, of course, later Christian cultures in Europe.

A recent find indicates that the spiritual use of cannabis was known by some. In 2020, a team of archaeologists found direct evidence for the spiritual use of cannabis in the ancient Judahite religion. These researchers discovered the psychoactive chemical (THC), found only in cannabis, within burned residues on a temple altar in the ancient kingdom of Judah, dating to *c.* 700 BC.[53] For context, the first Jewish temple in Jerusalem was built approximately 950 BC, but it was not until after 650 BC that the Jerusalem temple was made the exclusive temple.

This is not the only evidence for cannabis use in ancient Mosaic religious practices. A passage in Exodus indicates the use of קנה בשם (*qaneh bosem*) in the formula for the production of the sacred anointing oil for priests.[54] For thousands of years, *qaneh bosem* has been translated as 'sweet cane' (or calamus), yet it is just as likely to be a Hebrew transliteration of a foreign word just like another ingredient in the holy oil, קנמון *qinamon*, or

---

[50] J. M. McPartland and W. Hegman, 'Cannabis utilization and diffusion patterns in prehistoric Europe: A critical analysis of archaeological evidence', *Vegetation History and Archaeobotany*, 27(4) (2018), 627–634.

[51] M. Ren, Z. Tang, X. Wu, R. Spengler, H. Jiang et al., 'The origins of cannabis smoking: Chemical residue evidence from the first millennium BCE in the Pamirs', *Science Advances*, 5(6) (2019), article eaaw1391.

[52] The Yamnaya Culture, a proto-Indoeuropean culture, may have introduced the practice of smoking cannabis by 3500 BC. See T. Long, M. Wagner, D. Demske, C. Leipe, and P. E. Tarasov, 'Cannabis in Eurasia: Origin of human use and Bronze Age trans-continental connections', *Vegetation History and Archaeobotany*, 26(2) (2017), 245–258.

[53] E. Arie, B. Rosen, and D. Namdar, 'Cannabis and frankincense at the Judahite Shrine of Arad', *Tel Aviv*, 47(1) (2020), 5–28.

[54] 'Take the following fine spices: 500 shekels of liquid myrrh, half as much (that is, 250 shekels) of fragrant cinnamon, 250 shekels of fragrant calamus [קנה בשם *qaneh bosem*], 500 shekels of cassia – all according to the sanctuary shekel – and a hin of olive oil. Make these into a sacred anointing oil, a fragrant blend, the work of a perfumer. It will be the sacred anointing oil.' NIV, Exodus 30:23–25.

cinnamon.[55] Just as the Hebrew *qinamon* can be linked to the Greek *kinnamonon*, it seems the Hebrew *qaneh bosem* can be linked to the Greek *kánnabis*. This idea was originally proposed by Polish anthropologist Sara Benetowa in 1936, yet it fell into obscurity. One obvious reason for this obscurity is the intense cultural taboo around cannabis: 1936 was the same year as the release of the anti-cannabis propaganda film *Reefer Madness*.[56] There can be little doubt that, in the modern era, cultural taboos have actively suppressed knowledge regarding entheogens. This suppression also seems to have occurred in the ancient past, as well. An example of such a taboo can even be seen in Exodus, which demands a punishment of exile to any common person making use of a similar formula:

> Say to the Israelites, 'This is to be my sacred anointing oil for the generations to come. Do not pour it on anyone else's body and do not make any other oil using the same formula. It is sacred, and you are to consider it sacred. Whoever makes perfume like it and puts it on anyone other than a priest must be cut off from their people.'[57]

Why is it so common for cultures to institute taboos on the use of entheogenic plants and mushrooms? On the one hand, it may be that psychoactive plants and fungi are dangerous – cultures that suppress them may be healthier. Yet, ancient Greece and Roman sources regularly discussed more toxic psychoactive plants, like henbane or datura.[58] So, alternatively, the taboos might be motivated by the transformative nature of entheogenic experiences, insofar as they pose a disruptive threat to dominant cultural power structures. Power structures, particularly those dealing with magic and spirituality, may have been more likely to sustain their power if they forbade the common use of entheogenic plants. That is, if only the high priests could access entheogenic experiences, lesser priests and common people were less likely to be having different ideas. Yet, it is still a mystery why entheogens were virtually unknown during most of

---

[55] J. Benner, 'The facts about kaneh bosem' (Ancient Hebrew Research Center, 2020), www.ancient-hebrew.org/studies-words/facts-about-kaneh-bosem.htm.

[56] M. Booth, *Cannabis: A History* (London: Macmillan, 2015).

[57] NIV, Exodus 30:31–33.

[58] K. Fatur, '"Hexing herbs" in ethnobotanical perspective: A historical review of the uses of anticholinergic Solanaceae plants in Europe', *Economic Botany*, 74(2) (2020), 140–158.

European history and so virulently opposed during the 'Drug War' of the last 50 years. Now that the taboos are lifting on entheogen use,[59] it is possible to scientifically study the effects of these powerful sources of neuroplasticity and variation in the imagination and ask the following question: how will they affect the future evolution of the conscious mind?

Sympathy and the Replication of Conscious Experience

Now we turn to the second evolutionary mechanism in conscious experience: sympathy. Variations in experience, like variations in the genetic code, are a crucial driver of evolution. But how can these variations in conscious experience be reproduced between people?[60] A great deal of thinking has considered the outsize role of language and imitation in replicating experiences and human culture. We propose that sympathy is an even more primitive mechanism for spreading experiences between people. Sympathy, or fellow feeling, seemingly facilitates the reproduction of conscious feelings and affective states of mind. Sympathetic resonance between people allows the mental experiences of one person to be shared with another.[61]

In 1976, Richard Dawkins introduced 'memes' – as the counterpart to genes – as the 'units of imitation' that might explain cultural evolution.[62] Darwin himself had originally proposed that imitation played a major role in man's rapid cultural evolution:

> If some one man in a tribe, more sagacious than the others, invented a new snare or weapon, or other means of attack or defence, the plainest self-interest, without the assistance of much reasoning power, would prompt the other members to imitate him; and all would thus profit.[63]

[59] For one of many examples, the Entheogenic Plant and Fungus Policy Act of 2020 has decriminalised natural psychedelics including magic mushrooms, ayahuasca, and mescaline in the jurisdiction of Washington, DC.

[60] Individual memory is too vast a subject to discuss here. Yet, in Donald Hebb's *The Organization of Behavior: A Neuropsychological Theory* (New York, NY: Wiley, 1947) the 'mnemonic trace' is described as 'a lasting pattern of reverberatory activity'.

[61] A comprehensive review of resonance in physics, human experience, and design is provided by the authors in Lomas et al., 'Resonance as a design strategy'.

[62] R. Dawkins, *The Selfish Gene*, 40th anniversary edn (Oxford: Oxford University Press, 2016).

[63] C. Darwin, *The Descent of Man, and Selection in Relation to Sex*, 2 vols. (New York, NY: Appleton, 1871).

While imitation is valuable for accumulating and spreading cultural knowledge, it is a cognitive capacity that is largely unique to humans. Some groups of chimpanzees can transmit simple skills from parent to child – for instance, cracking nuts with a stone and anvil – but their process of social imitation is extremely slow. Scientists have found that it takes *multiple years* for young chimpanzees to learn nut cracking skills.[64]

If imitation is largely unique to humans, what came before? Darwin suspected that a social nature needed to evolve before imitation. In another sweet concordance with Wallace,[65] Darwin claimed (in *The Descent of Man*) that a key factor of sociality was *sympathy*:

> In order that ... the apelike progenitors of man should become social, they must have acquired the same instinctive feelings ... they would have felt uneasy when separated from their comrades, for whom they would have felt some degree of love ... all this implies some degree of sympathy.

Both Darwin and Wallace viewed sympathy as a primitive and essential factor in human social evolution. Thus, we claim that, even before the development of language or imitation (imitation is specifically necessary for 'memetic' cultural evolution),[66] sympathy provided a mechanism for reproducing conscious feelings. To understand how, the term *sympathy* needs some historical framing, especially since its modern meaning has shifted to mean something more like 'pity'.

Darwin was familiar with the Scottish economist Adam Smith, who, in his first book, investigated the role of sympathy in human society.[67] Smith

[64] Further, recent work disputes that social 'copying' is, in fact, the mechanism for the transmission. See E. Bandini, J. Grossmann, M. Funk, A. Albiach-Serrano, and C. Tennie, 'Naïve orangutans (*Pongo abelii and Pongo pygmaeus*) individually acquire nut-cracking using hammer tools', *American Journal of Primatology*, 83(9) (2021), article e23304.

[65] 'When the social and sympathetic feelings came into active operation, and the intellectual and moral faculties became fairly developed, man would cease to be influenced by "natural selection" in his physical form ... [but] his mind would become subject to those very influences from which his body had escaped', A. R. Wallace, 'The origin of human races and the antiquity of man deduced from the theory of "natural selection"', *Journal of the Anthropological Society of London*, 2 (1864), clviii–clxxxvii.

[66] S. Blackmore, 'Imitation and the definition of a meme', *Journal of Memetics – Evolutionary Models of Information Transmission*, 2(11) (1998), 159–170.

[67] A. Smith, *The Theory of Moral Sentiments* (London: Printed for A. Miller and A. Kincaid and J. Bell in Edinburgh, 1759).

explains how sympathy causes the automatic exchange of feelings, such that people feel good when other people feel good and feel bad when other people feel bad. Smith claimed that this sympathetic capacity gives people a natural incentive for moral action: that is, it feels good to make others feel good.

Today, the sympathy of Darwin or Wallace might be described with the term 'empathy'. Yet, empathy is a rather new word, introduced in 1909 as a translation of the German *Einfühlung* – a term used by Theodor Lipps and other psychologists to describe the *psychische Resonanz* that occurred during aesthetic experiences.[68] In a modern context, empathy is sometimes described as 'an effortful process by which we try to comprehend another's experience' whereas sympathy is 'a direct perceptual awareness of another person's experience akin to the phenomenon of sympathetic resonance'.[69] Thus, we aim to understand human sympathy through the physical concept of sympathetic resonance.

Resonance, Mirror Neurons, and Rhythmic Entrainment

The idea of sympathetic resonance is one of the most powerful and generalisable ideas in the physical sciences. It occurs when 'a system – a physical oscillator – is subjected to a periodic driving force by an external agency'.[70] Even in ancient times, people may have noticed how beating a drum can make another drum rattle or hum, even at a distance. The Greek Stoic philosophers taught that the universe was in sympathetic resonance with itself.[71] Today, modern cosmological models use sympathetic resonance to explain the origin of the universe during the inflationary stage of the 'Big Bang'.[72]

---

[68] S. Lanzoni, 'Sympathy in mind (1876–1900)', *Journal of the History of Ideas*, 70(2) (2009), 265–287.

[69] J. Decety and T. Chaminade, 'Neural correlates of feeling sympathy', *Neuropsychologia*, 41(2) (2003), 127–138.

[70] A. P. French, *Vibrations and Waves* (New York, NY: MIT Press, 1966).

[71] '[Chrysippus] first holds that the whole is unified by *pneuma*, which pervades it completely, and by which the universe is held together and stabilised and is sympathetic with itself', Alexander of Aphrodisias, *On Mixture*, 216.14–16; 'Meditate often on the interconnectedness and mutual interdependence of all things in the universe. For in a sense, all things are mutually woven together and therefore have an affinity for each other – for one thing follows after another according to their tension of movement, their sympathetic stirrings, and the unity of all substance', Marcus Aurelius, *Meditations*, 6.38.

[72] The winners of the 2002 Dirac Prize for physics claim that, during cosmic inflation, an oscillating 'inflaton' field resonated with all the other fields in the

While sympathetic resonance is a firmly scientific concept, it also played a central role in the history of magic.[73] And, perhaps because resonance seemed so magical, it took hundreds of years for scientists to accept the concept of resonance as more than a metaphor, namely as a common principle operating across all physical media, in acoustic systems, electromagnetic systems, gravitational systems, etc.[74] It is also common, however, to invoke the concept of 'resonance' to explain properties and values within human social systems, e.g., describing an artwork that 'resonates' with a viewer or a politician that 'resonates' with his or her followers.[75] Popular business books even provide guidance on how to give speeches or presentations to *resonate* with an audience.[76] Is this sort of human resonance a metaphor, or is it a mechanism? Consider the view of neurologist Oliver Sacks:

> People sing together and dance together in every culture, and one can imagine them having done so around the first fires, a hundred thousand years ago ... In such a situation, there seems to be an actual binding of nervous systems accomplished by rhythm.[77]

Some human resonances surely occur in the brain due to the oscillatory nature of neurons and neural circuitry. The brain contains vast hierarchies of rhythmic, reverberatory electrical oscillations; with all these oscillators, the brain inevitably resonates with itself and with stimuli in the environment. One simple and strong resonance effect, measurable via EEG, occurs when perceiving rhythmic pulses of light.[78] Oddly, some

standard model, producing particles. The Big Bang, in effect, created everything in the universe through this resonance. See L. Kofman, A. Linde, and A. A. Starobinsky, 'Towards the theory of reheating after inflation', *Physical Review D*, 56(6) (1997), article 3258. Special thanks to Patrick Cooper for drawing this to our attention.

[73] 'How do magic spells work? By sympathy, and by the natural concord of things that are alike and opposition of things that are different', Plotinus (b. 204 AD), *The Enneads*, IV. 4. 40.

[74] M. Buchanan, 'Going into resonance', *Nature Physics*, 15 (2019), article 203.

[75] H. Rosa, *Resonance: A Sociology of Our Relationship to the World* (Cambridge, MA: Polity Press, 2019).

[76] N. Duarte, *Resonate: Present Visual Stories That Transform Audiences* (Hoboken, NJ: Wiley, 2013).

[77] O. Sacks, *Musicophilia: Tales of Music and the Brain* (New York, NY: Knopf, 2007).

[78] C. S. Herrmann, 'Human EEG responses to 1–100 Hz flicker: Resonance phenomena in visual cortex and their potential correlation to cognitive phenomena', *Experimental Brain Research*, 137(3–4) (2001), 346–353.

of the most powerful neural resonances are correlated with the induction of hallucinations with geometrical patterns similar to those seen during psychedelic altered states.[79]

A more complex form of human resonance is known as 'motor resonance'. This phenomenon describes resonances between individuals, and it occurs when observations of human actions produce matching activity in the observer's brain and behaviour. For instance, an observer who watches someone pick up an egg will have, in their brain, a similar set of neural firing patterns as when they pick up an egg themselves. Observation of another person's actions triggers associated intentions, goals, or affective states; these resonant associations allow individuals to gain inference on the person they are observing. Astonishingly, scientists have measured this motor resonance – it occurs not just when observing actions directly but also when simply listening to stories describing actions or when reading stories about actions.[80]

We propose that the mechanism of sympathetic resonance enables the sharing of *feelings* (or at least some approximation of the feelings) between people during interpersonal interactions or through their interactions with artefacts. The proposed neural-cognitive basis for this sympathetic resonance is the so-called 'mirror neuron system'.[81] In a now famous essay, Ramachandran claimed that advances in the mirror neuron system seemed to be the key mechanism underpinning humanity's 'great leap forward' in evolutionary success:

> Anytime you watch someone else doing something (or even starting to do something), the corresponding mirror neuron might fire in your brain, thereby allowing you to 'read' and understand another's intentions, and thus to develop a sophisticated 'theory of other minds'.[82]

---

[79] B. C. Ter Meulen, D. Tavy, and B. C. Jacobs, 'From stroboscope to dream machine: A history of flicker-induced hallucinations', *European Neurology*, 62(5) (2009), 316–320; T. P. Cowan, 'Devils in the ink: William Burroughs, Brion Gysin, and geometry as a method for accessing intermediary beings', *Aries*, 19(2) (2019), 167–211.

[80] R. A. Zwaan, L. J. Taylor, and M. De Boer, 'Motor resonance as a function of narrative time: Further tests of the linguistic focus hypothesis', *Brain and Language*, 112(3) (2010), 143–149.

[81] G. Rizzolatti, L. Fadiga, L. Fogassi, and V. Gallese, 'Resonance behaviors and mirror neurons', *Archives italiennes de biologie*, 137(2) (1999), 85–100.

[82] V. S. Ramachandran, 'Mirror neurons and imitation learning as the driving force behind "the great leap forward" in human evolution', *Edge.Org* (2000), 1–7.

Long before language permitted the exchange of complex ideas, human sympathetic resonance seems to have enabled the exchange of conscious mental states. These experiences weren't shared with complete fidelity, of course, but the phenomenon might be comparable to how music provides a meaningfully similar reproduction of rich affective experiences between people. The ability to *feel* the music or to *feel* what another person is feeling is, of course, highly dependent on prior cultural experience. But, subject to cultural attunement, the rhythms of music do seem to 'induce affective states' through resonances with the naturally oscillatory brain:[83]

> The brain does not . . . 'compute' keys of melodic sequences, and it does not 'infer' meters of rhythmic input. Rather, it resonates to music [due to] oscillation of neural populations, rhythmic bursting, and neural synchrony.[84]

In a group of people, the sympathetic exchange of intersubjective emotional states is sometimes referred to as 'the vibe'. Because people have the ability to feel what other people in a group are feeling, the direct exchange of experience can lead to the emergence of prototypical group entrainment states that may evolve depending on their contribution to important cultural activities. For instance, certain group states could support vigour and enthusiasm during war or support calm focus for cognitive activities like complex tool manufacturing. This sympathetic resonance can also help explain what Émile Durkheim called 'collective consciousness' or 'collective effervescence' during tribal ceremonies:

> Once the individuals are gathered together, a sort of electricity is generated from their closeness and quickly launches them to an extraordinary height of exaltation . . . Thus the men of the clan and the things which are classified in it form by their union a solid system, all of whose parts are united and vibrate sympathetically.[85]

---

[83] W. J. Trost, C. Labbé, and D. Grandjean, 'Rhythmic entrainment as a musical affect induction mechanism', *Neuropsychologia*, 96 (2017), 96–110.

[84] E. W. Large, 'Neurodynamics of music', in M. R. Jones, R. R. Fay, and A. N. Popper (eds.), *Music Perception* (New York, NY: Springer, 2010), pp. 201–231.

[85] É. Durkheim, *The Elementary Forms of the Religious Life*, trans. J. Swain (London: George Allen & Unwin, 1912).

Rhythmic entrainment is a form of resonance activity that describes the spontaneous synchrony that occurs when two or more closely tuned oscillators interact with each other. Entrainment was discovered by the Dutch scientist Christiaan Huygens in 1665, after observing the emergent synchrony of the pendula of two clocks placed on a common beam. The capacity for rhythmic entrainment in music, dance, and other rituals may have played a significant role in human cultural evolution. In modern studies, entrainment has been found to improve cooperation in teams,[86] enhance social bonding, and increase prosocial behaviour.[87]

Although rhythmic entrainment naturally emerges in oscillatory systems throughout nature, 'holding a musical beat' is a uniquely human characteristic.[88] Although there are isolated instances in animals (like the YouTube star Snowball, a dancing parakeet),[89] these exceptions prove the rule.[90] Non-human animals are, for some reason, deeply resistant to the natural phenomena of sympathetic resonance or rhythmic entrainment.

Yet, animals *should* have the capacity for rhythmic entrainment – simply because much of everyday neural functioning relies upon entrainment. For instance, the rhythmic beating of the heart is driven by its rhythmic entrainment to central pattern generators in the spinal cord.[91] This example precisely shows why it can be dangerous for animals to allow entrainment to external rhythms – it is one thing for the heart of an animal to be entrained by the spinal cord, but it would be deadly if its heart could be entrained by a clever predator. Did animals evolve defence

---

[86] S. S. Wiltermuth and C. Heath, 'Synchrony and cooperation', *Psychological Science*, 20(1) (2009), 1–5.

[87] T. C. Rabinowitch and A. N. Meltzoff, 'Synchronized movement experience enhances peer cooperation in preschool children', *Journal of Experimental Child Psychology*, 160 (2017), 21–32.

[88] S. A. Kotz, A. Ravignani, and W. T. Fitch, 'The evolution of rhythm processing', *Trends in Cognitive Sciences*, 22 (2018), 896–910.

[89] A. D. Patel, J. R. Iverson, M. R. Bregman, and I. Schulz, 'Experimental evidence for synchronization to a musical beat in a nonhuman animal', *Current Biology*, 19 (10) (2009), 827–830.

[90] A. D. Patel, 'The evolutionary biology of musical rhythm: Was Darwin wrong?', *PLoS Biology*, 12(3) (2014), article e1001821.

[91] R. L. Calabrese, F. Nadim, and Ø. H. Olsen, 'Heartbeat control in the medicinal leech: A model system for understanding the origin, coordination, and modulation of rhythmic motor patterns', *Journal of Neurobiology*, 27(3) (1995), 390–402.

mechanisms to protect against external entrainment?[92] If so, perhaps humans evolved biological or cultural capacities to 'let down the guard' and open up to certain kinds of rhythmic entrainment with other people. Notably, some psychoactive plants and alcohol seem to enhance the perception of music, dance, and interpersonal sympathy; in ages past, perhaps they helped catalyse new depths of human resonance and rhythmic entrainment?

Our understanding of human resonance is rapidly advancing in response to various advances in biosensing. Hyperscanning is a brain imaging technique used to measure correlated brain activity between multiple people as they interact.[93] Scientists have found, for instance, that there is greater neural synchrony between the brains of two conversing friends than between the brains of two conversing strangers.[94] Neuroscientist Suzanne Dikker et al. discovered that brain synchrony is associated with the performance of students in classrooms and collaborators on teams.[95] Technologist Ramesh Rao applied hyperscanning to the heart beats of multiple people performing the highly repetitive movements in Kundalini meditation; his findings showed a fascinating pattern of emergent interpersonal coherence.[96] In the ritual context of fire walking, researchers have found evidence for physiological resonance (synchronisation in heart rate) between participants and related observers.[97] Thus, new scientific techniques are revealing how different

[92] For instance, the cuttlefish uses flickering skin patterns to mesmerise prey. See M. J. How, M. D. Norman, J. Finn, W.-S. Chung, and N. J. Marshall, 'Dynamic skin patterns in cephalopods', *Frontiers in Physiology*, 8 (2017), article 393.

[93] A. Czeszumski, S. Eustergerling, A. Lang, D. Menrath, M. Gerstenberger et al., 'Hyperscanning: A valid method to study neural inter-brain underpinnings of social interaction', *Frontiers in Human Neuroscience*, 14 (2020), article 39.

[94] S. Kinreich, A. Djalovski, L. Kraus, Y. Louzoun, and R. Feldman, 'Brain-to-brain synchrony during naturalistic social interactions', *Scientific Reports*, 7(1) (2017), 1–12.

[95] S. Dikker, L. Wan, I. Davidesco, L. Kaggen, M. Oostrik et al., 'Brain-to-brain synchrony tracks real-world dynamic group interactions in the classroom', *Current Biology*, 27(9) (2017), 1375–1380; D. A. Reinero, S. Dikker, and J. J. Van Bavel, 'Inter-brain synchrony in teams predicts collective performance', *Social Cognitive and Affective Neuroscience*, 16(1–2) (2021), 43–57.

[96] G. Quer, J. Daftari, and R. R. Rao, 'Heart rate wavelet coherence analysis to investigate group entrainment', *Pervasive Mobile Computing*, 28 (2016), 21–34.

[97] D. Xygalatas, 'The biosocial basis of collective effervescence: An experimental anthropological study of a fire-walking ritual', *Fieldwork in Religion*, 9(1) (2015), 53–67.

ritual practices can support synchrony, rhythmic entrainment, and physiological resonance. Because the synchrony of neural firing helps create conscious experiences in individuals, biological synchrony between people perhaps indicates the shared and extended nature of consciousness.[98] Many rituals seem to enable humans to share a collective consciousness.

The evolution of rhythmic rituals may have had a profound effect on social cohesion. War drums are used in many cultures to bind together groups of warriors and give them a shared mental state. However, not all rituals need to be rhythm-centred to induce the sympathetic replication of mental states. For instance, the extensive archaeological record of Mayan culture – and advances in epigraphy – have painted a picture of a cultural fabric held together through a framework of spirituality accessed via altered states of consciousness by all levels of society.[99] As interpreted by David Stuart, the acclaimed Mayanist credited with deciphering the Maya Codex, altered states were accessed through exhaustive repetitive dance, public incense burning, extreme fasting, entheogens, and, most prominently, the auto-sacrificial letting of blood by rulers (Figure 8.7).[100]

Bloodletting was resonant but it was not rhythmic. The high king was believed to be the 'vessel' of 'ku'h' – a force understood as the divine energy flowing within all things. Standing atop pyramids in front of their gathered subjects, kings spilled their blood in ritualised acts symbolising the transferal of the ku'h within them to the people, thus establishing their societal power. These were bloody ceremonies of pain, as the ruler would pierce their own tongue or penis with a holy instrument, such as an engraved stingray spine. Through the human capacity for sympathy, ritual spectacles like this would have produced extremely powerful shared experiences. Perhaps ritual experiences that resonate more will replicate more.

[98] A. L. Valencia and T. Froese, 'What binds us? Inter-brain neural synchronization and its implications for theories of human consciousness', *Neuroscience of Consciousness*, 2020(1) (2020), article niaa010.

[99] D. Stuart, 'Ideology and classic Maya kingship', in V. L. Scarborough (ed.), *A Catalyst for Ideas: Anthropological Archaeology and the Legacy of Douglas Schwartz* (Santa Fe, NM: School of American Research Press, 2005), pp. 257–286; S. Houston and D. Stuart, 'Of gods, glyphs and kings: Divinity and rulership among the classic Maya', *Antiquity*, 70(268) (1996), 289–312.

[100] S. Houston, 'Into the minds of ancients: Advances in Maya glyph studies', *Journal of World Prehistory*, 14(2) (2000), 121–201.

FIGURE 8.7 Mayan bloodletting ceremony, as depicted in Lintel 24 from Structure 23 at Yaxchilan (Am1923, Maud.4, The Yaxchilan Lintels, The British Museum). © The Trustees of the British Museum.

Resonance, in the context of this discussion, is a mechanism for *experience transmission* between people. In culture, as was exhibited by the Maya, the role of resonant experiences between minds may be integral to maintaining the coherence of societal structure. Yet it is challenging to translate an internal experience into a transmittable external action, a reality reflected in the Mayan glyph *tzak* meaning 'conjure (in the context of vision)', which shows a fish held in a left hand.[101] As Stuart points out, a slippery fish in hand may be a metaphor for the illusive nature of our

[101] Stuart, 'Ideology and classic Maya kingship'.

liminal and ineffable dance with the divine. Yet, the attempt to face the challenge of sharing internal experience is present in the rich variety of public ceremonies found throughout human culture.

Harmony as a Selection Pressure in Human Evolution

In this chapter, we have proposed *imagination* as a mechanism of mental variation and *sympathy* as a mechanism of mental reproduction. We now consider the classical concept of *harmony* as a selection mechanism within conscious experience. To begin, we address the following question: what is conscious experience? The neuroscientist Stanislaw Deheane put forward one of the most popular theories of consciousness, known as the 'Global Neuronal Workspace'.[102] Given that a person can only have one conscious experience at a time, Dehaene claims that consciousness consists of the mental elements that have temporarily won the ongoing competition for access to this unified space (that is, the global neuronal workspace). Thus, conscious experience is that which is integrated into the global neural workspace – and this competition for dominance in awareness serves as an evolutionary pressure.

Beyond the competition of individual mental elements, we suggest that the mind *harmonises* competing mental impulses into a single conscious experience. How might the classical concept of harmony help explain complex selection pressures in the mind? We hypothesise that harmonisation drives mental selection forces because coherence or periodic synchrony between mental elements is stabler and lower in free energy.[103] In the following section, we will show how harmonisation might be able to explain how the mind can integrate the diverse experiences of an individual

---

[102] S. Dehaene, H. Lau, and S. Kouider, 'What is consciousness, and could machines have it?', *Science*, 358(6362) (2017), 486–492.

[103] A similar discussion can be found in Safron's consideration of self-organising harmonic modes: A. Safron, 'An Integrated World Modeling Theory (IWMT) of consciousness: Combining integrated information and global neuronal workspace theories with the free energy principle and active inference framework; towards solving the Hard problem and characterizing' agentic causation', *Frontiers in Artificial Intelligence*, 3 (2020), article 30.

or collective consciousness. The humanistic concept of harmony can be found in both Eastern and Western classical philosophy, psychology, neuroscience, and physics, but we will begin with evolutionary theory.

Wallace, the younger co-discoverer of Darwinian evolution, presented 'harmony' as an overarching selection force in the evolution of life: 'health, strength, and long life are the results of a harmony between the individual and the universe that surrounds it'. For Wallace, a plant or animal is in harmony with its environment when it is well-matched to it. When the environment changes, a species can fall out of harmony. Natural selection can return an animal species to harmony with a changing universe, 'like the governor of a steam engine',[104] but only through slow change over many generations. In contrast, Wallace notes that 'man is kept in harmony by his mind'.[105]

In *The Descent of Man*, Darwin quotes Wallace, saying that 'man is enabled through his mental faculties to keep with an unchanged body in harmony with the changing universe'.[106] The concept of harmony here is clear enough: organisms that better harmonise with their environment will flourish. In the classical meaning,[107] harmony is defined as the joining together of diverse elements into a well-*fitted* whole: so, 'survival of the fittest' could also reasonably be termed 'survival of the harmonised'.

Harmony is an important concept in many cultures, including in indigenous spiritual systems in the Americas and in Africa. Historically, the concept emerged as a dominant philosophical theme almost simultaneously

---

[104] Wallace provides an early view on cybernetics or control theory: 'The action of this principle is exactly like that of the centrifugal governor of the steam engine, which checks and corrects any irregularities almost before they become evident; and in like manner no unbalanced deficiency in the animal kingdom can ever reach any conspicuous magnitude, because it would make itself felt at the very first step, by rendering existence difficult and extinction almost sure soon to follow', Alfred Russel Wallace, 'On the tendency of varieties to depart indefinitely from the original type' (Ternate, February 1858), ed. Charles H. Smith, http://people.wku.edu/charles.smith/wallace/S043.htm.

[105] Wallace, 'The origin of human races', 160 and 184.

[106] Darwin, *The Descent of Man*, Part 1, ch. 5, p. 158.

[107] This history is represented with astonishing clarity – and a direct call to revolution – in HRH the Prince of Wales, T. Juniper, and I. Skelly, *Harmony: A New Way of Looking at Our World* (London: Harper Collins, 2010). Also see Lomas and Xue, 'Harmony in Design', 5–64.

around 550 BC in ancient China[108] and in ancient Greece,[109] a coincidence perhaps comparable to the simultaneous discovery of Darwinism. While the concept of 'harmony' might strike some as unscientific, it actually played a key role in the development of the Western sciences.

The first known empirical philosophers in the Western tradition were the Greek Pythagoreans (*c. 500 BC*), who were known for their mathematical theory of harmony.[110] While previous thinkers had proposed that the world was fundamentally made of elements like air, water, fire, or aether, the Pythagoreans believed that the cosmos was made of *mathematical structure*. As a result, they believed that natural harmonies in mathematics should manifest themselves within the physical cosmos. This viewpoint is still provocative today, in part because it seems to integrate an immaterial or even spiritual perspective with scientific inquiry and mathematics.

Notably, the Pythagoreans tested their mathematical model of harmony in what appears to be the first documented scientific experiment. The historian Leonid Zhmud explains that 'Hippasus [*c. 530–450 BC*] conducted an experiment with bronze discs, confirming the numerical expressions of the principal concords discovered by Pythagoras, 2:1 for the octave, 3:2 for the fifth, and 4:3 for the fourth.'[111] The Pythagoreans empirically demonstrated that a bronze chime twice the thickness (ratio of 1:2) of another made a musical octave while a chime 150 per cent thicker (ratio of 2:3) played a musical fifth, and so on. These simple combinations of pure numbers were also found in the ratios used to tune stringed instruments. For instance, a guitar string pressed at half the length will play an octave higher, at a ratio of 2:1. These early

---

[108] In sixth-century BC China, Confucians and Daoists believed that harmony described the joining of diversity into wholeness. It wasn't the elimination of differences or 'sameness', it was a coherent integration of variety, like a soup with many contrasting ingredients, coming together as a whole. See C. Li, 'The philosophy of harmony in classical Confucianism', *Philosophy Compass*, 3(3) (2008), 423–435.

[109] Philolaus (470–385 BC), born 100 years after Pythagoras (*c. 570–495 BC*), was the first Pythagorean known to write about harmony: 'Harmony is generally the result of contraries: for it is the unity of multiplicity, and the agreement of discordances' (Nicomachus, *Arith. Intr.*, 2. 509, DK 10).

[110] The terms 'philosophy' and 'cosmos' are attributed to Pythagoras. See C. Riedweg, *Pythagoras: His Life, Teaching, and Influence* (Ithaca, NY: Cornell University Press, 2008).

[111] L. Zhmud, 'Aristoxenus and the Pythagoreans', in C. Huffman (ed.), *Aristoxenus of Tarentum: Discussion* (New Brunswick, NJ: Transaction, 2012), pp. 223–249.

experiments provided empirical evidence to support the Pythagorean hypothesis, that transcendent mathematical relationships govern the material universe and are manifested as literal harmonies in the physical cosmos as well as in the mind.[112]

Evidence from contemporary neuroscience suggests that the Pythagorean hypothesis may be worthy of some further consideration. For instance, the brain is not just rhythmic, it is also profoundly harmonic. Electrical brainwaves are structured by doublings of frequency – in music, these doublings are known as octaves. Just as an octave above middle A (440 Hz) is 880 Hz, each brainwave band is roughly double in frequency: gamma brainwaves are roughly double the frequency of beta waves, which are roughly double the frequency of alpha waves, and so on. These doublings support periodic synchrony; for instance, the peaks of one frequency's wave can 'kiss' the peaks of another frequency's wave every other cycle. In this manner, harmonic ratios support periodic synchrony and coupling between neural circuits. Because irrational ratios can *prevent* synchrony, famous irrational ratios like the golden mean have been proposed as mechanisms useful for preventing interference between different neural functions.[113] For example, functions at 'high theta' won't interfere with functions at 'low theta' if the high and low frequencies are separated by a ratio of the golden mean.

The harmonic structure of the brain can be described as a nested hierarchy of frequency couplings. By analogy, consider how the rapid play of a violin can fit within the slower frequencies of the musical pulse indicated by a conductor's baton: in the same way, rapid, localised ~40 Hz gamma cycles can fit (or nest) into the brain-wide metrical structure of ~4 Hz theta oscillations. Thus, local and global frequencies can be coupled together through a nested hierarchy of periodic phase synchrony.[114]

[112] J. Godwin, *The Harmony of the Spheres: A Sourcebook of the Pythagorean Tradition in Music* (Rochester, NY: Inner Traditions International, 1993); A. Balbi, *The Music of the Big Bang: The Cosmic Microwave Background and the New Cosmology* (Berlin and London: Springer Science & Business Media, 2008).

[113] B. Pletzer, H. Kerschbaum, and W. Klimesch, 'When frequencies never synchronize: The golden mean and the resting EEG', *Brain Research*, 1335 (2010), 91–102.

[114] P. Uhlhaas, G. Pipa, B. Lima, L. Melloni, S. Neuenschwander et al., 'Neural synchrony in cortical networks: History, concept and current status', *Frontiers in Integrative Neuroscience*, 3 (2009), article 17.

When different neural circuits fire in synchrony, they are perceived together in consciousness;[115] long-range synchronisation is understood as the binding force for the perception of smaller 'nested' elements. For instance, neural synchrony in the imagination blends and synthesises different concepts into new concepts.[116] Oscillatory chunks of language are merged together through rhythmic nested synchrony – phonemes nested in syllables nested in words nested in phrases and so on.[117]

It is this dynamic, hierarchical *harmony* – the symphony of the brain – that seems to generate the integrated unified nature of conscious experience.[118] As neuroscientists seek a 'common currency'[119] to explain the link between immaterial mental experiences and physical brain activities, the concept of harmony is a strong candidate for this common currency. Scientists have even discovered that electrical oscillations in the brain can produce so-called 'harmonic modes':[120] electrical patterns resembling the beautiful patterns of sand that can emerge on a vibrating Chladni plate. Science is only beginning to unravel how the brain's massive hierarchy of neural oscillation fits together in a rhythmic, metrical structure. While much remains to be discovered, mathematical principles of harmony really do seem to be manifested in the brain. Pythagoreans can be proud.

But even if principles of harmony operate in the brain and mind, what makes harmony a selection pressure? A guitar string offers a simple illustration. When a guitar string is plucked, a massive range of initial frequencies are generated, but only the *fit* will survive;[121] that is, only the

---

[115] A. K. Engel and W. Singer, 'Temporal binding and the neural correlates of sensory awareness', *Trends in Cognitive Sciences*, 5(1) (2001), 16–25.

[116] A. Vyshedskiy and R. Dunn, 'Mental synthesis involves the synchronization of independent neuronal ensembles', *Research Ideas and Outcomes*, 1 (2015), article e7642.

[117] G. Buzsáki, *The Brain from Inside Out* (New York, NY: Oxford University Press, 2019), ch. 6.

[118] F. Varela, J.-P. Lachaux, E. Rodriguez, and J. Martinerie, 'The brainweb: Phase synchronization and large-scale integration', *Nature Reviews Neuroscience*, 2(4) (2001), 229–239.

[119] G. Northoff, S. Wainio-Theberge, and K. Evers, 'Is temporo-spatial dynamics the "common currency" of brain and mind? In quest of "Spatiotemporal Neuroscience"', *Physics of Life Reviews*, 33 (2019), 34–54.

[120] S. Atasoy, I. Donnelly, and J. Pearson, 'Human brain networks function in connectome-specific harmonic waves', *Nature Communications*, 7(1) (2016), 1–10.

[121] D. A. Jaffe and J. O. Smith, 'Extensions of the Karplus–Strong plucked-string algorithm', *Computer Music Journal*, 7(2) (1983), 56–69.

vibrations with wavelengths literally *fitting* the length of the string will be amplified. All other frequencies will be damped. These frequencies are the string's fundamental frequency and its harmonics. Thus, a 'survival of the harmonised' process operates; from among a multitude of frequencies, harmony selects only those that fit the length of the string. Notably, it is this selection process that gives rise to the overall sound of a guitar. By direct analogy, rhythmic elements in the brain that do not fit or harmonise with dominant neural rhythms will be unable to enter conscious awareness.[122] In this way, we hypothesise that harmony serves as a 'fitness function' for the brain: neural circuits are tuned through selection forces to maximise harmony.

We have been speaking of harmony in the brain, but what about harmony in human experience? In everyday mental experiences, harmony offers a powerful and intuitive description of several deep and vital human motivations. For instance, harmony is deeply linked to 'happiness' as an encompassing driver of human experience. When researchers gathered open-ended definitions of happiness from nearly 3,000 adults (from Argentina, Brazil, Croatia, Hungary, India, Italy, Mexico, New Zealand, Norway, Portugal, South Africa, and the United States), they found that 'inner harmony' was the most common everyday 'lay-person' definition of happiness.[123] A harmonious state of mind is indeed a pleasant one – yet we also need tension and unpleasant 'cognitive dissonance' to motivate actions that return us to states of greater harmony.[124] After all, the best music relies on dissonance and the best stories rely on narrative tension; harmony appears to be greater when there are discords to harmonise or challenges to overcome.

There may be more extreme states of mental harmony, as well. Mihaly Csikszentmihalyi describes the trance-like peak experience of top performers, the so-called flow states, using the concept of harmony: 'Flow is

---

[122] P. Fries, 'Rhythms for cognition: Communication through coherence', *Neuron*, 88 (1) (2015), 220–235; R. F. Helfrich, A. Breska, and R. T. Knight, 'Neural entrainment and network resonance in support of top-down guided attention', *Current Opinion in Psychology*, 29 (2019), 82–89.

[123] A. Delle Fave, I. Brdar, M. P. Wissing, U. Araujo, A. Castro Solano et al., 'Lay definitions of happiness across nations: The primacy of inner harmony and relational connectedness', *Frontiers in Psychology*, 7 (2016), 30.

[124] B. Weiner, *Human Motivation* (London: Psychology Press, 2013), p. 305.

the way people describe their state of mind when consciousness is harmoniously ordered.'[125] Flow states don't occur when one is distracted or when one's attention is split; they occur only when a performer interacts *whole-mindedly* with the demands of their environment. When there is a dynamic match between a person's skill and the demands of the task, this can result in a state of harmony or 'flow'. An example could be conjured in the image of a skilled musician entranced in an effortless union of mind and sound, where thought is not needed to determine placement of fingers upon a fretboard – rather, the music seems to emerge from a direct link between the mind and the instrument.

Extending the analogy of the musician, a group flow or harmony in improvisational music ('jamming') depends upon a harmonious balance of listening, feeling, and generation by each individual musician. When a group harmony is achieved, the intersection of diverse individual musical ideas is profoundly unified. New ideas may need to resonate in order to reproduce, but it is through selection-by-harmony that they can evolve to fit within the context of a larger social system.

Harmony does not only occur in the brain or within individual experiences, it also manifests itself as the social harmony that binds people together in community. Individual actions that fit the expectations of a larger social group are less likely to be dampened and extinguished. Some behaviours harmonise with the rhythms of other people, and some don't. People seek interpersonal harmony with each other because it feels good – and because alignment into friendships and larger social groups generally supports individual health and reproductive success. The attunement of people within social groups appears to be necessary for joint attention,[126] coordinated action, and shared intention. Harmony, as a selection pressure in the evolution of the social mind, may have enabled more collaborative behaviours and larger social groups. Groups of humans that don't harmonise will have more conflict and fragment over time, whereas

---

[125] M. Csikszentmihalyi, *Flow: The Psychology of Optimal Experience* (New York, NY: Harper & Row, 1990).

[126] M. Tomasello, M. Carpenter, J. Call, T. Beyne, and H. Moll, 'Understanding and sharing intentions: The origins of cultural cognition', *Behavioral and Brain Sciences*, 28(5) (2005), 675–691.

groups that harmonise well may grow ever larger. Over time, then, harmonious groups may come to dominate the less harmonious.

One critique of harmony is that it isn't a scientific concept. Yet, during the European Enlightenment, the notion of harmony and scientific humanism were closely aligned. Alexander von Humboldt, the world explorer and inspiration to Darwin, named his masterpiece of science *Kosmos* as a reference to the Pythagorean philosophy. Similarly, both Copernicus and Newton credited Pythagoreans for their greatest insights.[127] Other scientists, like Galileo, Kepler, Descartes, Spinoza, and Hooke (among many others) scientifically investigated harmony. In the first years of the early Royal Society, for instance, there were dozens of experiments conducted on the science of harmony.[128]

For a variety of esoteric reasons,[129] the concept of harmony has largely fallen out of consideration in the sciences. Even Karl Popper was critical of it.[130] Richard Dawkins referred to the idea of ecological harmony between species as 'dotty mysticism'.[131] Perhaps. But, if this classical concept motivates people to seek inner harmony, to design a more harmonious global society,[132] and to pursue a more lasting harmony with nature, perhaps it should be understood better through authentic scientific inquiry:

> Nature itself is a harmonic system ... if we ignore the principles that sustain that harmony, nature's essential balance and equilibrium become quite literally disordered.[133]

---

[127] J. E. McGuire and P. M. Rattansi, 'Newton and the "Pipes of Pan"', *Notes and Records of the Royal Society of London*, 21(2) (1966), 108–143.

[128] P. Gouk, *Music, Science and Natural Magic in Seventeenth-Century England* (New Haven, CT: Yale University Press, 1999).

[129] W. J. Hanegraaff, *Esotericism and the Academy: Rejected Knowledge in Western Culture* (Cambridge: Cambridge University Press, 2012).

[130] But perhaps due to a misunderstanding, as discussed in C. Li and D. Düring, 'Harmony: Origin of totalitarianism or patron of pluralism', *Journal of East–West Thought*, 10(2) (2020), 1–9.

[131] R. Dawkins and Y. Wong, *The Ancestor's Tale: A Pilgrimage to the Dawn of Life* (London: Hachette, 2016).

[132] D. Mahiet, 'Rethinking harmony in international relations', *Journal of International Political Theory*, 17(3) (2021), 257–275.

[133] HRH the Prince of Wales, Juniper and Skelly, *Harmony*.

Synopsis

Do evolutionary forces operate within the human mind? In this chapter, we have examined factors in conscious experience that support variation, reproduction, and selection: the key ingredients for natural selection. We propose that *imagination* provides a key source of *variety* in human experience. We propose that *sympathy* serves as a powerful mechanism for the *replication* of conscious experiences. Finally, we propose that principles of *harmony* govern the *selection* of mental experiences, both within individuals and within collective consciousness.

Looking across the world today, we see the results of a mere ~70,000 years of cultural adaptation from the moment our modern human ancestors left Africa to populate the lands beyond. A band of possibly fewer than 1,000 individuals crossed the Red Sea strait and began the lasting migration wave that would disperse humanity over mountains, across oceans, and onto every corner of habitable land that this planet has to offer. In the generations that followed, many conditions led to a diverse set of human experiences that influenced a seemingly endless variety of experiments in culture. Empires rose and fell, societies formed and dissolved, and ideas were born and transformed through it all. What we are left with is an archaeological library of human imagination.

As a species, we will always need to create new ways of thinking in order to adapt to our continuously changing realities. Now, in a time where globalisation and the information revolution have dissolved geographical boundaries of knowledge exchange, there may be a rational and appropriate need to reflect on our longstanding relationship with the boundaries of the mind, to expand our imagination of the universe and to seek a more harmonious future for ourselves within it.

Further scientific discovery can reveal how resonance and harmony shape our universe and how they relate to each other and to other basic concepts like entropy and the flow of free energy. Understanding these forces will surely inform the continued evolution of humanity. We are humbled by the complexity of the emergence of new realities within the mind – the birth of new ideas remains a mystery. The enigma of our imagination may be as elusive as the origin of life; by exploring this enigma, we seek the spirit of human inspiration. From the shaman's drum in Mongolia to the sakau stones in Micronesia, we have as a species been driven to evolve as explorers of the mind. By continuing to redefine the boundaries of our conscious experience, we can forge and formulate reality itself.

Epilogue: The Continuing Frontier

In the field of biomimetics, lessons from nature are used to inform engineering. How might we build upon the evolution of cultural rituals and cognitive technologies to inform the design of new and transformative experiences today?

A. Lin personal account – In late March of 2019, during the spring equinox, I was invited to a private installation deep within the Yucatán jungles of Mexico. It was designed by the landscape artist James Turrell and built by conservationists/explorers Roberto Hernández and Claudia Madrazo Hernández with a philosophical grounding in Maya cosmology. It was described to me as a technology to facilitate connection to the cosmos and the binding forces in nature. The installation consisted of a massive stone pyramid built over an underground *cenote* (a naturally formed limestone cavern deep enough to access groundwater). The interior of the pyramid framed a series of internal spaces, each designed to challenge and 'unlock the mind'.

Embedded at the centre of the base layer was a voluminous space that, due to the lack of corners, gave the illusion of having no end. From no specific source, a perfectly ubiquitous light shifted in colour almost imperceptibly. My hosts referred to this as 'The Ganzfeld' (Figure 8.8). They instructed me to sit quietly and stare into the seemingly endless abyss – an entrancing, amorphous canvas of formless space with nothing in it but us. Then, we took a small series of stairs (Figure 8.9) to a half-spherical space referred to as 'The Metasphere'. We lay on our backs on top of a circular platform and stared into the dimly glowing dome above. Suddenly, the sea of diffuse light began to oscillate and flicker from alternating hidden sources. As I stared into the pulsing featureless space, I saw morphing multicolour images emerge like a dance of constructive and destructive interference patterns in my mind. I heard an audio signal at a similar alternating frequency – it pulsed and

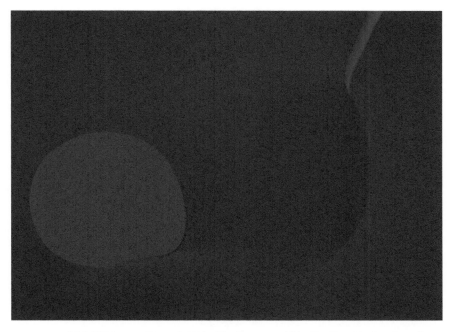

FIGURE 8.8 'Ganz field' inside the pyramid in the Yucatán jungles of Mexico (designed by James Turrell; built by Roberto Hernández and Claudia Madrazo Hernández). Photo: Albert Lin.

hummed and seemed to entangle itself in my mental images. I was told that the frequencies were attuned to disrupt the normal rhythmic operation of my brain.

When we reached the third level, we saw a dark and cavernous room holding the 'skyspace' – a starkly cut elliptical view of the sky that was framed sharply on all sides by pulsing diffuse lights filling the interior of the space (Figure 8.10). A final minimal staircase rose to the opening, giving way to a reflection pool situated on top of the pyramid (Figure 8.11). From the top of the steps, we emerged from the interior to witness the endless jungle horizon. Even the colours and sounds of the jungle seemed significantly brighter upon emerging from the pyramid. We sat just below the skyspace as the sky tone shifted with the setting sun. The boundaries of both colour and depth played unexplainable games with my perception.

FIGURE 8.9 Channel through the heart of the pyramid into the cenote.
Photo: Albert Lin.

FIGURE 8.10 The 'Skyspace' inside the pyramid (the grey oval is open sky). Photo: Albert Lin.

FIGURE 8.11 Reflection pool on the top level of the pyramid. Photo: Albert Lin.

FIGURE 8.12 Light hitting the cenote surface within the pyramid. Photo: Albert Lin.

A single long channel led from the top of the pyramid down several hundred feet to the dark blue waters of the natural *cenote* below (Figure 8.12). I will never forget those few minutes during the spring equinox when a stunning beam of sunlight travelled through the incense-filled heart of the pyramid and exploded into the rippling, clear waters.

Over the course of three days, my experience there was punctuated only by the rhythms of ceremony, the simple ritual of eating small meals and our ongoing discussion of the cosmos and our relationship to it. Without any entheogenic substances, this otherworldly and existential experience – generated through architecture, immersive light, and conversation – was enveloping to the point of mystical ego-death.

This experience shows how the design of new cognitive technologies might tap into the more existential aspects of the human mind. The frontiers of this very human quest will surely be advanced by many explorers of art, technology, science, and philosophy. With a greater understanding of imagination, sympathetic resonance, and harmony, how might we envision new designs that could enlighten our experience and aid our evolution as a more resilient and harmonious species?

The Earth has never been more connected together at a global scale. Yet the world can feel, at times, profoundly disconnected. We have seemingly infinite access to information and experience – all the great works of science, literature, music, film, and more – yet it is rare to participate in the physical, embodied group ritual experiences that have generally held societies together in the past. Dancing together is powerful but, in 2021, it is deeply provocative. As humans – if we have the capacity – do we also have the *responsibility* to imagine new ceremonies, rituals, or technologies that could help bind together our future societies? We leave this discussion wondering what resonant experiences might enable humans to better harmonise with each other and with the ever-changing natural world. Most of all, we continue to wonder, where will our collective imagination take us next?

Acknowledgements

We would like to thank Ramesh Rao, Vilayanur Subramanian Ramachandran, Sheldon Brown, Johan Reinhard, Timothy Furnish, Fadel Zeidan, Adam Halberstad, Mark Geyer, Cassandra Vieten, Patrick Colman, Mark Meyers, Tim Kelly, Eliah Aronoff-Spencer, Thomas Garrison, Lee Stein, Roberto and Claudia Hernández, Clara Wu, Wade Davis, Leo Trottier, Tanner Cusick, Deborah Forester, Adam Safron, Tim Mullen, Patrick Cooper, Michael Skirpan, Don Norman, P. J. Stappers, Paul Hekkert, Caiseal Beardow, Emily Cross, Paul Stamets, Harry Diakoff, Julika Lomas, Tommy Cohen, and Pascal Gagneux for their intellectual contribution, guidance, and support in our explorations. We would also like to thank Emily Joan Ward, Janet Gibson, and Andrew Fabian for the honour of being invited to present our work in this esteemed lecture series. We dedicate this chapter to our respective fathers, Douglas Lin and Robert Lomas, who inspired us to explore and contemplate the furthest realms of our imagination, and thus our universe.

Index

Index

Index

Index